PILOT'S POCKET DECODER

Christopher J. Abbe

McGraw-Hill
New York San Francisco Washington, D.C. Auckland Bogotá Caracas
Lisbon London Madrid Mexico City Milan Montreal New Delhi
San Juan Singapore Sydney Tokyo Toronto

Library of Congress Cataloging-in-Publication Data
Abbe, Christopher J.
Pilot's pocket decoder / Christopher J. Abbe
p. cm.
ISBN 0-07-007549-2 (alk. paper)
1. Aeronautics—Abbreviations. 2. Aeronautics—Terminology.
3. English language—Terms and phrases. I. Title.
TL509.A18 1998
629.13'01'4—DC21 98-4210
CIP

McGraw-Hill

A Division of The ***McGraw-Hill*** *Companies*

8 9 10 11 12 13 DOC/DOC 0 9 8 7 6 5 4

ISBN 0-07-007549-2

The sponsoring editor for this book was Shelley Ingram Carr, the editing supervisor was Patricia V. Amoroso, and the production supervisor was Clare B. Stanley. It was set in Times Ten by Jaclyn J. Boone and Michele Betterman of McGraw-Hill's Professional Book Group in Hightstown, N.J.,

Printed and bound by R. R. Donnelley & Sons Company.

Contents

Introduction

This book has been in preparation for many years. By the time I had completed college, advanced flight ratings, certificates, and airline training, I had thousands of initials and abbreviations that were needed the first day of my private pilot ground school.

This book is compiled mostly from what was written on black, blue, and green chalkboards and from a few class handouts. These classes include global navigation, fuel and lubricants, turbines, aircraft reciprocating power plants, meteorology, aircraft electronics, aircraft airframes, aviation safety, numerous ground schools, and many other aviation seminars and classes devoted to similar topics. The book contains initials either commonly used in the cockpit, in ground training, or in stories from fellow pilots. I hope this book will be helpful throughout your future as an aviator or aviation enthusiast.

There are a few things you should know to help in your search for the unknown meaning behind the mysterious letters in this book. The first and largest section of the book is the abbreviations and initials chapter. Several chapters dealing with more specific categories follow. Each chapter is in alphabetical order to make reference simple. Start your search in Chapter One. If the meaning is not found there, check the weather-related chapter. If you still can't find the meaning, consider the possible context of the abbreviation/initials and search the relevant chapter alphabetically. If you are unsuccessful after checking all chapters of this book, then try logistics. Most abbreviations are created by removing vowels and leaving only the consonants. Although many of the abbreviations and initials have both general and weather-related meanings, they aren't listed in both chapters. Therefore, if you don't find it in one chapter, try another. Abbreviations and initials have been placed in the chapter that deals with their most common use.

When you find your abbreviation/initials, look at how they are printed. If all the letters are capitals, then they are initials. In other words, the first letter of each main word is used to make the short form of common reference.

Example:

AFD **A**irport **F**acility **D**irectory

If the letters are all in lowercase, they form an abbreviation for a word.

Example:

dsplcd **displaced**

There are a few exceptions to this format. If it is a single-letter abbreviation, it is also a capital.

Example:

I **I**njected

Abbreviations commonly written in capital letters (such as on charts or in directories) appear that way in the book.

Example:

RGT TFC **R**ight **T**raffic

Months also have their first letter capitalized.

Example:

Sep **S**eptember

The weather chapter is in capital letters because that is how most weather reports are printed.

Example:

LTGICCCCG **L**ightnin**g** **I**n **C**louds, **C**loud to **C**loud, and **C**loud to **G**round

If there is more than one meaning for an abbreviation, the meanings appear on the same line.

Example:

mod **mod**ulate, **mod**ulation

With the initials, if there is more than one meaning, each meaning is on a separate line.

Example:

DA **D**ecision **A**ltitude
DA **D**ensity **A**ltitude

This is because abbreviations are only one word and require less space. Initials are several words and require their own line.

This book does not include all clubs, organizations, associations, companies, and so forth. Only the most common ones are listed. Under no circumstances should this book replace instruction from a flight instructor.

Acknowledgments

There are many I would like to thank for their help with the completion of this book. They include my family, friends, and instructors. Special thanks are in order to my father, James Abbe. He is a pilot and my mentor for my career choice, and he encouraged me to compile this book. This would still be on notebook paper without his help.

Mary Reymann was my instructor for my CFI, CFII, and MEI. I would like to thank her for quality instruction on how to be a flight instructor.

Thanks also to Dr. David Rouleau, Terry Michmerhuizen, and David Schneider for their help in formatting and proofreading the manuscript. I appreciate their help and encouragement with the book and with my career.

I owe a great deal of gratitude to Bob Munley, Walter Musial, William Julian, Dr. David Rouleau, Bill Bethau, Julie Clippard, Mary Reymann—my flight instructors and the instructors at Western Michigan University School of Aviation Science. Thanks also to Ward Seeley, Ben Gibson, Joe Mosteller, Dr. Sheila Smith, Kevin Raymond, and John DeGroot.

ONE

Abbreviations and Initials

A	Alert
A	Class "A" airspace
A&E	Airframe and Engine
A&ME	Aeronautical and Mechanical Engineering
A&P	Airframe and Power plant
A/A	Air to Air
A/B	Afterburner
A/C	Aircraft
A/C	Approach Control
A/F	Air-to-Fuel ratio
A/FD	Airport/Facility Directory
A/G	Air to Ground
A/H	Altitude or Height
A/P	Autopilot
A/S	Airspeed
AAAE	American Association of Airport Executives

CHAPTER ONE

AADC	Approach And Departure Control
AAF	Army Air Force
AAI	Angle of Approach Indicator
AAI	Arrival Aircraft Interval
AAL	Above Aerodrome Level
AAS	Advanced Automation System
AAS	Airport Advisory Service
AAS	American Astronautical Society
AB	Air Base
abate	abatement
abbr	abbreviation
ABC	After Bottom Center
ABDC	After Bottom Dead Center
abm	abeam
abnml	abnormal
abs	absolute
AC	Advisory Circular
AC	Aerodynamic Center
AC	Air Carrier
AC	Air Conditioning
AC	Alternating Current
AC Form	Airman Certification Form

Abbreviations and Initials

ACARS	Aircraft Communications Addressing and Reporting System
ACAS	Airborne Collision Avoidance System
ACC	Active Clearance Control
ACC	Air Coordinating Committee
ACC	Area Control Center
ACCC	Area Control Computer Complex
accel	accelerate
access	accessory
accum	accumulate, accumulation
ACDO	Air Carrier District Office
ACE	Aviation Career Education
acft	aircraft
ACI	Airport Council International
ACLS	Automatic Carrier Landing System
ACLT	Actual Calculated Landing Time
ACN	Airborne Classification Number
ACO	Aircraft Certification Office
ACP	Airlift Command Post
ACP	Aniline Cloud Point
ACR	Air Carrier

ACR	Air Corps Reserve
act	actual
ACT GS	Actual Ground Speed
actuat	actuating
actvt	activate
actvty	activity
AD	Airworthiness Directive
AD	Apparent Drift
ADAS	Aircraft Departure and Arrival Sequencing
ADCUS	Advise Customs
ADF	Automatic Direction Finder
ADI	Attitude Director Indicator
ADIZ	Air Defense Identification Zone
adj	adjacent, adjustable
ADLY	Arrival Delay
adm	administration
ADM	Aeronautical Decision Making
ADMA	Aviation Distributors and Manufacturers Association
ADO	Airport District Office
ADR	Advisory Route

Abbreviations and Initials

ADS	Air Data System
adv	advisory
AEEC	Airline Electronic Engineering Committee
AEF	American Expeditionary Forces
AEG	Aircraft Evaluation Group
aero	aerodrome, aerodynamic, aeronautical
AESOP	Aircraft, Environment, Situation, Operation, and Personnel
AF&E	Airframe and Engine
AFA	Air Force Association
AFB	Air Force Base
AFCEA	Armed Forces Communications and Electronics Association
AFD	Airport Facility Directory
AFFF	Aqueous Film Forming Foam
AFI	African-Indian Ocean region (ICAO)
AFI	Authorized Flight Instructor
AFIS	Aerodrome Flight Information Service
AFL	Above Field Level

AFM	Aircraft Flight Manual
AFOVRN	Air Force Overrun
AFR	Actual Fuel Remaining
AFSC	Aggregate Friction Seal Coat
AFSS	Automated Flight Service Station
AFTI	Advanced Fighter Technology Integration
AGB	Accessory Gearbox
AGI	Advanced Ground Instructor
AGI	Authorized Ground Instructor
AGL	Above Ground Level
agri	agriculture
AH	Artificial Horizon
AHS	American Helicopter Society
AI	Attitude Indicator
AIA	Aircraft Industries Association of America
AIAA	American Institute of Aeronautics and Astronautics
AIF	Instrument Flight Instructor—Airplane (added rating)
ail	aileron
AIM	Airman's Information Manual

Abbreviations and Initials

AIP	Aeronautical Information Publication
AIP	Airport Improvement Program
AIRMET	Airman's Meteorological information
airpl	airplane
AIS	Aircraft Integrating System
AIT	Auto Ignition Temperature
AIT	Automated Information Transfer
ALA	Authorized Landing Area
alc	alcohol
ALERFA	Alert Phase
ALF	Auxiliary Landing Field
ALNOT	Alert Notice
ALPA	Airline Pilots Association
ALS	Approach Light System
ALSF-I	Approach Light System with sequenced Flashing lights
ALSF-II	Approach Light System with sequenced Flashing lights and red side row bars the last 1000 ft
ALSTG	Altimeter Setting
alt	alternate, altimeter

altd	altitude
altm	altimeter
ALTRV	Altitude Reservation
alum	aluminum
AM	Amplitude Modulation
AMA	Area Minimum Altitude
AMASS	Airport Movement Area Safety System
amdt	amendment
AME	Aviation Medical Examiner
AMEL	Airplane Multi-Engine Land
AMES	Airplane Multi-Engine Sea
AMM	Aircraft Maintenance Manual
AMOS	Automatic Meteorological Observing Station
amp	ampere
AMSL	Absolute Mean Sea Level
AMT	Aviation Maintenance Technician
AMVER	Automated Mutual-assistance Vessel Rescue system
AN	Air Force-Navy
ANDB	Air Navigation Development Board

Abbreviations and Initials

ANG	Air National Guard
ANGB	Air National Guard Base
ANPRM	Advanced Notice of Proposed Rule Making
ANSI	American National Standards institute
ant	antenna
anten	antenna
AOA	Angle Of Attack
AOE	Airport Of Entry
AOPA	Aircraft Owners and Pilots Association
ap	approach, autopilot
APA	Allied Pilots Association
APAPI	Abbreviated Precision Approach Path Indicator
apc	approach
APC	Area Positive Control
APFA	Association of Professional Flight Attendants
APFD	Autopilot Flight Director
API	American Petroleum Institution
APL	Airport Lights

app	approach
APP CON	Approach Control
APPM	Accident Prevention Program Manager
approp	appropriate
approx	approximately
Apr	April
apt	airport
APU	Auxiliary Power Unit
AR	Aspect Ratio
ARAC	Army Radar Approach Control
ARAC	Aviation Rulemaking Advisory Committee
ARB	Air Reserve Base
ARC	Aeronautical Research Council
ARFF	Aircraft Rescue Firefighting/equipment
ARINC	Aeronautical Radio Incorporated (discrete frequency)
arngmt	arrangement
ARO	Airport Reservation Office
ARP	Airport Reference Point
arpk	airpark

arpt	airport
arr	arrival, arrive
arrest	arresting
ARS	American Rocket Society
ARSA	Airport Radar Service Area
ARSA	Aeronautical Repair Station Association
ARSR	Air Route Surveillance Radar
ARTAS	Air traffic control Radar Tracking And Server
ARTCC	Air Route Traffic Control Center
ARTS	Automated Radar Terminal System
ASAP	As Soon As Possible
ASD	Aircraft Situation Display
ASDA	Accelerate—Stop Distance Available
ASDE	Airport Surface Detection Equipment
ASEL	Airplane Single-Engine Land
ASES	Airplane Single-Engine Sea
ASF	Additional Safety Factor
ASF	Air Safety Foundation
ASI	Aviation Safety Inspector

ASI	Airspeed Indicator
ASL	Above Sea Level
ASLAR	Aircraft Surge Launch And Recovery
ASME	American Society of Mechanical Engineers
ASOS	Automated Surface Observing System
ASP	Arrival Sequencing Program
asph	asphalt
ASR	Airport Surveillance Radar
ASR	Altimeter Setting Region
assn	association
asst	assistant
ASTA	Airport Surface Traffic Automation
ASTM	American Society for Testing Materials
ATA	Actual Time of Arrival
ATA	Air Transportation Association of America
ATA	Airport Traffic Area
ATA	Aviation Training Association

ATAC Air Transport Association of Canada

ATC After Top Center

ATC Air Traffic Control

ATCAA Air Traffic Control Assigned Airspace

ATCCC Air Traffic Control Command Center

ATCO Air Taxi Commercial Operator

ATCRBS Air Traffic Control Radar Beacon System

ATCSCC Air Traffic Control System Command Center

ATCT Airport Traffic Control Tower

ATD Actual Time of Departure

ATDC After Top Dead Center

ATE Actual Time En route

ATE Automatic Test Equipment

ATEC Aviation Technician Education Council

ATF Aerodrome Traffic Frequency

ATFM Air Traffic Flow Management

ATHODYD	Aerothermodynamic Duct
ATIS	Automatic Terminal Information Service
ATLAS	Abbreviated Test Language for Avionics Systems
ATM	Aerodynamic Twisting Moment
ATM	Air Traffic Management
atmos	atmosphere
atnd	attended
atndt	attendant
ATP	Airline Transport Pilot
ATR	Air Transport Radio
ATS	Aeronautical Testing Services
ATS	Air Traffic Services
att	attended
ATZ	Aerodrome Traffic Zone
Aug	August
AUSA	Association of the United States Army
auth	authorized
auto	automatic
AUTOB	Automatic Observing station

AUW	All-Up Weight
aux	auxiliary
AVASI	Abbreviated Visual Approach Slope Indicator
avbl	available
AVCOMM	Aviation Communications
avd	avoid
AVGAS	Aviation Gasoline
avia	aviation
avn	aviation
AVN	Aviation Standards National field office
avt	active
avtn	aviation
aw	airworthiness
AW&ST	Aviation Week and Space Technology
AWG	American Wire Gage
AWIPS	Advanced Weather Interactive Processing System
AWOS	Automatic Weather Observing/reporting System
AWR	Airborne Weather Radar

AWW	Severe Weather forecast Alert
az	azimuth
azm	azimuth

B	Battery
B/CA	Business and Commercial Aviation
bag	baggage
bal	balance
bat	battery
BBC	Before Bottom Center
BBDC	Before Bottom Dead Center
BC	Back Course
BC	Bearing Change
BC	Bottom Center
BC	British Columbia
BCAR	British Civil Airworthiness Requirements
BCM	Back Course Marker
bcn	beacon
BCOB	Broken Clouds Or Better
bcst	broadcast
BDC	Bottom Dead Center

bdcst	broadcast
bdry	boundary
BERM	Snowbank(s) containing earth/gravel
BFO	Beat Frequency Oscillator
BFR	Biennial Flight Review
BGI	Basic Ground Instructor
BHP	Brake Horsepower
BIH	International time bureau
BITE	Built-In Test Equipment
BLC	Boundary Layer Control
bldg	building
blkd	bulkhead
blks	blocks
bln	balloon
BM	Back Marker
BMEP	Brake Mean Effective Pressure
bnd	bound
BOW	Basic Operating Weight
BPR	Bypass Ratio
brg	bearing
brkt	bracket
brz	bronze

BS	Broadcast Station
BSFC	Brake Specific Fuel Consumption
BTC	Before Top Center
BTDC	Before Top Dead Center
BTL	Beacon Tracking Level
BTU	British Thermal Unit
bush	bushing
BVOR	Broadcast VOR
bynd	beyond

C	Centrifugal force
C	Circling
C/A	Course/Acquisition
C/S	Call Sign
C/W	Continuous Wave
CA	Closing Angle
CAA	Civil Aeronautics Administration
CAA	Civil Aeronautics Authority
CAAS	Class "A" Airspace
CAATS	Canadian Automated Air Traffic System
CAB	Civil Aeronautics Board

cad	cadmium
CADC	Central Air Data Computer
CADIZ	Canadian Air Defense Identification Zone
CAFE	Corporate Average Fuel Economy
calc	calculated
CAM	Civil Aeronautics Manual
CAMI	Civil Aeromedical Institute
can	canceled
cap	captain
CAP	Civil Air Patrol
capt	captain
capy	capacity
CAR	Caribbean region (ICAO)
CAR	Civil Airworthiness Regulation
carb	carburetor
CARF	Central Altitude Reservation Function
CAS	Calibrated Airspeed
CAS	Collision Avoidance System
CAS	Computed Airspeed
CAT	Carburetor Air Temperature

cat	category
CB	Circuit Breaker
CBAS	Class "B" Airspace
CBD	Center of Business District
CBSA	Class "B" Surface Area
CC	Chart Convergency
CC	Compass Course
CC	Courtesy Car
CCA	Continental Control Area
CCAS	Class "C" Airspace
CCSA	Class "C" Surface Area
ccw	counterclockwise
CD	Chart Distance
CD	Clearance Delivery
CDAS	Class "D" Airspace
CDI	Course Deviation Indicator
CDP	Compressor Discharge Pressure
CDSA	Class "D" Surface Area
CDT	Central Daylight Time
CDT	Compressor Discharge Temperature
CDT	Control Departure Time
CDU	Control Display Unit

CEAS	Class "E" Airspace
ceil	ceiling
CENRAP	Center Radar Arts Presentation/Processing
CEP	Central East Pacific
CERAP	Combined Center—RAPCON
cert	certificate
CESA	Class "E" Surface Area
CFA	Controlled Firing Area
CFCF	Central Flow Control Function
CFD	Computational Fluid Dynamics
CFI	Certificated Flight Instructor
CFIA	Certificated Flight Instructor—Airplane
CFII	Certificated Flight Instructor—Instrument
cfm	confirm
CFM	Cubic Feet per Minute
CFR	Call For Release
CFR	Code of Federal Regulations
CFRP	Carbonfibre-Reinforced Plastics
CFS	Cubic Feet per Second
CFWSU	Central Flow Weather Service Unit

CG	Center of Gravity
CGAS	Class "G" Airspace
CGAS	Coast Guard Air Station
CGL	Circling Guidance Lights
ch	channel
CH	Compass Heading
CH	Critical Height
chan	channel
chng	change
chrg	charge
chrgr	charger
chrt	chart
CHT	Cylinder Head Temperature
CIP	Compressor Inlet Pressure
circum	circumference
CIT	Combustor Inlet Temperature
CIT	Compressor Inlet Temperature
civ	civil
CL	Centerline Lights
CL	Clearance Limit
CL	Condition Lever
CLC	Course Line Computer

clkws	clockwise
clnc	clearance
CLR	Command, Leadership, and Resource management
clsd	closed
CLT	Calculated Landing Time
CM	Chrome-Molybdenum
CMNPS	Canadian Minimum Navigation Performance Specification
cmplt	complete
cmpltn	completion
cmsd	commissioned
cmsn	commission
cmsnd	commissioned
CN	Change Notice
CN	Compass North
cncld	canceled
cnl	cancel
CNS	Central Nervous System
CNS	Communications, Navigation, and Surveillance
cntrln	centerline
co	county

COC	Cone Of Confusion
COCO	Coordinator Of Chain Operations
coef	coefficient
com	commercial, communications
COMET	Continental U.S. Meteorological Teletype system
coml	commercial
COMLO	Compass Locator
comm	communications
comp	composite
compr	compression
compt	compartment
comsnd	commissioned
con	control
conc	concrete
conduc	conductor
conn	connecting
const	construction
Cont	Continental (engine manufacturer)
cont	contour, control
coord	coordinates
COP	Change-Over Point

copter	helicopter
corr	corridor
CP	Center of Pressure
CP	Circular Polarization
CP	Command Post
CPC	Constant Pressure Chart
CPL	Current Flight Plan
CPR	Compressor discharge (Pressure) Ratio
CPU	Central Processing Unit
CR	Counter-Rotating
CR	Current Regulator
CRM	Cockpit Resource Management
CRM	Crew Resource Management
CRP	Compulsory Reporting Point
CRS	Certified Repair Station
crs	course
CRT	Cathode Ray Tube
CS/T	Combined Station and Tower
CSFC	Cruise Specific Fuel Consumption
CST	Central Standard Time
cstg	casting

CTA	Control Area
CTA	Controlled Time of Arrival
CTAF	Common Traffic Advisory Frequency
ctc	contact
CTE	Cruise Thermal Efficiency
ctl	control
CTLZ	Control Zone
CTM	Centrifugal Twisting Moment
ctr	center
cu	cubic
CVFR	Controlled VFR
cw	clockwise
CW	Continuous Wave
CWA	Center Weather Advisory
CWSU	Center Weather Service Unit
CWY	Clearway
cyl	cylinder
CZ	Control Zone
D	Day
D	Dual-wheel (landing gear)

D	Dynamic
D/F	Ground Direction finding
DA	Decision Altitude
DA	Density Altitude
DA	Drift Angle
DAIR	Direct Altitude and Identity Readout
DALR	Dry Adiabatic Lapse Rate
DAMI	Designated Airworthiness Maintenance Inspection
DAR	Designated Airworthiness Representative
dat	date
db	decibel
dbl	double
DC	Direct Current
DC	Douglas Corporation
DC	Dry Chemical
dcmsn	decommission
dcmsnd	decommissioned
dct	direct
DDT	Double Dual-Tandem (landing gear)
Dec	December

del	delivery
demo	demonstration
demsn	decommission
dep	depart, department, departure
dept	department
DER	Designated Engineering Representative
dev	deviation
DEWIZ	Distant Early Warning Identification Zone
DF	Direction Finder
DFCS	Digital Flight Control System
DFSV	Pilot Forecaster Service
DG	Directional Gyro
DGAC	Direction Generale de l' Aviation Civile
DGI	Directional Gyroscopic Indicator
DH	Damage History
DH	Decision Height
DI	Direction Indicator
dia	diameter
DINS	Digital Inertial Navigational System

dir	direction
dis	distance
disabld	disabled
displ	displaced
dist	distance
distrb	distributor
DL	Direct Line
dlad	delayed
dlt	delete
dltd	deleted
dly	daily
DMA	Decision-Making Aids
DMA	Defense Mapping Agency
DME	Distance Measuring Equipment
DME/P	Precision Distance Measuring Equipment
DMEA	Damage Modes and Effects Analysis
DMIR	Designated Manufacturing Inspection Representative
dmstn	demonstration
dn	down

DOA	Department Of Aviation
DOC	Department Of Commerce
DOC	Designated Operational Coverage
DOD	Department Of Defense
dom	domestic
DOT	Department Of Transportation
dp	departure
DPCR	Departure Procedure
DR	Dead Reckoning
DRC	Direct-Reading Compass
drct	direct
drftd	drifted
DSC	Differential Scanning Calorimetry
DSP	Departure Sequencing Program
dsplc	displaced
dsplcd	displaced
dsr	desired
dstc	distance
DT	Delay Time
DT	Displaced Threshold
DT	Dual-Tandem (landing gear)
DTC	Desired Track

DUAT	Direct User Access Terminal
dup	duplicate
DURGD	During Descent
durn	duration
DVA	Diverse Vector Area
DVFR	Defense Visual Flight Rules
DVOR	Doppler VOR
E	East
E-MSAW	En route Minimum Safe Altitude Warning
E/S	Engine Speed
E6-B	Flight Computer
ea	each
EAA	Experimental Aircraft Association
EAC	Expect Approach Clearance
EAC	Expect Approach Control
EADI	Electronic Attitude Direction Indicator
EAEC	European Airline Electronic Committee

EARTS	En route Automated Radar Tracking System
EAS	Equivalent Airspeed
EAT	Expected Approach Time
ebnd	eastbound
EC	Earth Convergency
ECCM	Electronic Counter-Countermeasures
ECM	Electronic Counter-Measures
ECM	Extra Crew Member
ECS	Environmental Control System
ECU	Electronic Control Unit
ED	Earth Distance
EDCT	Expected Departure Clearance Time
EDT	Eastern Daylight Time
EE	Electrical Engineer
EEC	Electronic Engine Control
EET	Estimated Elapsed Time
EFA	Eurofighter Aircraft
EFAS	En route Flight Advisory Service
EFC	Expect Further Clearance

eff	effective
effy	efficiency
EFIS	Electronic Flight Management System
EFR	Estimated Fuel Remaining
EGT	Exhaust Gas Temperature
EHF	Extremely High Frequency
EHP	Equivalent Horsepower
EHSI	Electronic Horizontal Situation Indicator
EICAS	Engine Indication and Crew Alerting System
EKW	Equivalent Kilowatts
EL	Electroluminescence
elec	electric
elev	elevate, elevation
ELR	Environmental temperature Lapse Rate
ELT	Emergency Locator Transmitter
emerg	emergency
EMF	Electromagnetic Force
EMI	Electromagnetic Interference
emp	empennage

enc	encoder
eng	engine
ent	entry
EOBT	Estimated Off Block Time
EOW	Empty Operating Weight
EP	Effective Pitch
EP	Extreme Pressure
EPA	Environmental Protection Agency
EPR	Engine Pressure Ratio
EPS	Engineered Performance Standards
EPT	Effective Performance Time
EPU	Emergency Power Unit
eq	equal
equip	equipment
equiv	equivalent
ER	Earth Rotation
ERAA	European Regional Airline Association
ESD	Electrostatic Discharge
ESFC	Equivalent Specific Fuel Consumption
ESHP	Equivalent Shaft Horsepower

ESP	En route Spacing Program
EST	Eastern Standard Time
ETA	Estimated Time of Arrival
ETAS	Effective True Airspeed
ETD	Estimated Time of Departure
ETE	Estimated Time En route
ETHP	Equivalent Thrust Horsepower
ETOPS	Extended Operations
ETP	Equal Time Point
ETR	Estimated Time of Return
EUR	European region (ICAO)
EVC	Engine Vane Control
exh	exhaust
exp	expires
exper	experimental
extd	extend, extended
exten	extended, extension
exting	extinguishing
F	Field
F	Sequenced Flashing lights
F/A	Fuel-to-Air ratio

FA	Final Approach
FA	Flight Attendant
FAA	Federal Aviation Administration
FAAP	Federal Aid to Airports Program
fac	facility
FAC	Final Approach Course
FADEC	Full-Authority Digital Electronic Control
FAF	Final Approach Fix
FAI	Federation Aeronautique Internationale
FAN	Federal Aviation News
FAP	Final Approach Point
FAPA	Future Aviation Professionals of America
FAR	Federal Aviation Regulation
FAS	Final Approach Segment
FAWS	Flight Advisory Weather Service
FBI	Federal Bureau of Investigation
FBO	Fixed Base Operator
FBW	Fly By Wire
FCC	Federal Communication Commission

FCLT	Freeze Calculated Landing Time
FCP	Final Control Point
FCU	Fuel Control Unit
FD	Flight Director
FDC	Flight Data Center (NOTAMs)
FDU	Flow Divider Unit
FE	Field Elevation
FE	Flight Engineer
Feb	February
FEFI/ TAFI	Flight Environmental Fault Isolation/ Turnaround Fault Isolation system
FEX	Flight Engineer written Exam
FF	Fuel Flow
FG	Fixed Gear
FI	Fuel Injected
FIA	Flight Instructor—Airplane
FIC	Fleet Installed Cost
FIC	Flight Information Center
FIDO	Flight Inspection District Office
FIFO	Flight Inspection Field Office
FIH	Instrument Flight Instructor—rotorcraft/Helicopter

FII	Instrument Flight Instructor—airplane
FIR	Flight Information Region
FIRC	Flight Instructor Refresher Clinic
FIS	Flight Information Service
FL	Flight Level
flex	flexible
flg	flashing
FLIFO	Flight Information
FLIP	Flight Information Publication
flt	flight
FM	Fan Marker
FM	Frequency Modulation
FMA	Final Monitor Aid
FMCS	Flight Management Computer System
FMS	Flight Management System
fnt	front
FO	First Officer
FOD	Foreign Object Damage
FOEB	Flight Operations Evaluation Board
FOI	Fundamentals Of Instruction

fone	telephone
FPL	Flight Plan
FPS	Feet Per Second
FRC	Full Route Clearance
FREM	Factory Remanufactured
freq	frequency
FRH	Fly Runway Heading
FS	Factor of Safety
FSDO	Flight Standards District Office
FSF	Flight Safety Foundation
FSPC	Fleet Spares Cost
FSPD	Freeze Speed Parameter
FSS	Flight Service Station
FSUC	Fleet Support Cost
ft	feet
FTD	Flight Training Device
FTL	Food, Transportation, and Lodging
FTP	Federal Test Procedure
fus	fuselage
FWCS	Flight Watch Control Stations
FX	Foreign Exchange

G	Generator
G	Gravitational force
G	Ground
G/S	Ground Speed
GA	General Aviation
GAAP	General Aviation Action Plan
GADO	General Aviation District Office
gal	gallon
GAMA	General Aviation Manufacturers Association
GASIL	General Aviation Safety Information Leaflet
GC	Ground Control
GCA	Ground Control Approach
GEM	Graphic Engine Monitor
gen	generator
genl	general
geo	geographic
GFE	Government Furnished Equipment
GLONASS	Global Orbiting Navigation Satellite System
GM	Greenwich Meridian

GMT	Greenwich Mean Time
GNC	Global Navigation Chart
GNSS	Global Navigation Satellite System
GOES	Geostationary Operational Environmental Satellite
govt	government
GP	Geometric Pitch
GP	Glidepath
GPH	Gallons Per Hour
GPI	Ground Point of Interception
GPM	Gallons Per Minute
GPO	Government Printing Office
GPS	Gallons Per Second
GPS	Global Positioning System
GPU	Ground Power Unit
GPWS	Ground Proximity Warning System
GR	Glide Ratio
gr	gross
GRI	Group Repetition Interval
grnd	ground
grvd	grooved
grvl	gravel

GS	Glideslope
GS	Ground Speed
GSAR	Great Southern Air Race
GSIP	Glideslope Intercept Point
GW	Gross Weight
GWT	Gross Weight
gyro	gyroscope
H	High
H	High altitude
H	Without voice
H/W	Headwind
HAA	Height Above Airport
HAI	Helicopter Association International
HAL	Height Above Landing
HAT	Height Above Touchdown
haz	hazard
HCL	Horizontal Component of Lift
HCW	Horizontal Component of Weight
HDF	High-frequency Direction Finder
hdg	heading
hel	helicopter

heli	heliport
HERF	High-Energy Radio Frequency
HF	High Frequency
HFR	High-Frequency Range
HIALS	High-Intensity Approach Light System
HIBAL	High-altitude Balloon
HIF	Instrument Flight Instructor – rotorcraft/Helicopter (added rating)
HIRL	High-Intensity Runway Lights
HIWAS	Hazardous In-flight Weather Advisory Service
HLD	Head-Level Display
HMD	Helmet-Mounted Display
HMU	Hydromechanical Unit
HMV	Heavy Maintenance Visit
hol	holiday
horiz	horizontal
HOTAS	Hands On Throttle And Stick
hp	horsepower
HP	Heated Pitot
HP	High Performance

HP	Holding Pattern
HP	Hot Prop
HPC	High-Pressure Compressor
HPT	High-Pressure Turbine
HPZ	Helicopter Protection Zone
hr	hear, here, hour
HRA	Horizontal Resolution Advisory
HRD	High-Rate Discharge
hs	hours
HSCT	High-Speed Civil Transport
HSCT	Hypersonic Civil Transport
HSI	Horizontal Situation Indicator
HST	High-Speed Taxiway
HTTB	High-Technology Test Bed
HTZ	Helicopter Traffic Zone
HUD	Head-Up Display
HVOR	High VOR
HWD	Horizontal Weather Depiction charts
hyd	hydraulic
HySTP	Hypersonic System Technology Program
Hz	hertz

I	Injected
I	Island
IA	Indicated Airspeed
IA	Indicated Altitude
IA	Inspection Authorization
IA	Intercept Angle
IAC	Instrument Approach Chart
IAM	International Association of Machinists
IACA	International Air Charter Association
IAF	Initial Approach Fix
IAL	Instrument Approach and Landing chart
IAOPA	International council of Aircraft Owners and Pilots Association
IAP	Instrument Approach Procedure
IAS	Indicated Airspeed
IAS	Instrument Rating—Airplane (added rating)
IATA	International Air Transport Association
IBN	Identification Beacon

ibnd	inbound
ic	icing
IC	Integrated Circuits
IC	Internal Combustion
ICAO	International Civil Aviation Organization
ICC	Instrument Competency Check
ICC	Interstate Commerce Commission
ICO	Idle Cutoff
id	identification
ID	Inside Diameter
ident	identification, indentifier, identify
IE	Industrial Engineer
IEEE	Institute of Electrical and Electronics Engineers
IF	Intermediate Fix
IFALPA	International Federation of Airline Pilots Association
IFF	Identification Friend or Foe
IFIM	International Flight Information Manual
IFP	Instrument Rating—Foreign Pilot
IFR	Instrument Flight Rules

IFSS	International Flight Service Station
IGE	In-Ground Effect
IGI	Instrument Ground Instructor
ign	ignition
IGS	Instrument Guidance System
IGV	Inlet Guide Vanes
IH	Intercept Heading
IHS	Instrument Rating—rotorcraft/Helicopter (added rating)
illus	illustration
ILS	Instrument Landing System
IM	Inner Marker
IMC	Instrument Meteorological Conditions
IMO	International Meteorological Organization
IMTA	Intensive Military Training Area
in	inch, inches
IN	International Notices
inbd	inboard, inbound
INCERFA	Uncertainity phase

indic	indication, indicator
inop	inoperative
INP	If Not Possible
INREQ	Information Request
ins	inches
INS	Inertial Navigation System
insp	inspect, inspection
inst	instrument
int	intersection
intcp	intercept
inter	interior
intl	international
INVOF	In Vicinity Of
IOAT	Indicated Outside Air Temperature
IPC	Illustrated Parts Catalog
IPM	Illustrated Parts Manual
IPT	Intermediate Pressure Turbine
IR	IFR military training Routes
IR	Instrument Reference
IRA	Instrument Rating—Airplane
ireg	irregular

IRH	Instrument Rating—rotorcraft/Helicopter
IRS	Inertial Reference System
IRVR	Instrument Runway Visual Range
is	islands
ISA	International Standard Atmosphere
ISMLS	Interim Standard Microwave Landing System
ISO	International Standards Organization
ISP	Interim Support Plan
ISSS	Initial Sector Suite System
ITT	Interturbine Temperature
IVRS	Interim Voice Response System
IVSI	Inertial Lead or Instantaneous Vertical Speed Indicator
J	Jet
J-AID	Jeppesen Airport and Information Directory
Jan	January
JAR	Joint Airworthiness Requirements

JATO	Jet Assisted Take-Off
Jly	July
JNC	Jet Navigational Chart
Jne	June
JPATS	Joint Primary Aircraft Training System
junct	junction

K	Knot
K-Ice	Known Icing
KE	Kinetic Energy
kgs	kilograms
kHz	kilohertz
KI	Known Icing
KIAS	Knots Indicated Airspeed
km	kilometer
KMH	Kilometers per Hour
kn	knot
KOL	Kinds of Operations List
kol	kollsman
kt	knot
KTAS	Knots True Airspeed

kts	knots
kw	kilowatts
L	Left
L	Lift
L	Locator
L	Low
L	Low altitude
L/D	Lift/Drag ratio
L/F	Load Factor
L/MF	Low/Medium Frequency
LAA	Local Airport Advisory
LAAS	Low-Altitude Alert System
LAHSO	Land And Hold Short Operations
lat	latitude
LAWRS	Limited Aviation Weather Reporting Station
lb	pound
LBCM	Locator Back Course Marker
LBM	Locator Back Marker
lbs	pounds
LC	Landing Chart

LC	Local Call
LC	Local Control
LCB	Line of Constant Bearing
LCC	Life Cycle Cost
LCC	Loran-C navigational Chart
LCD	Liquid Crystal Display
LCG	Load Classification Group
LCN	Load Classification Number
lctd	located
lctr	locator
LDA	Localizer type Directional Aid
LDAZ	Landing Distance Available
ldg	landing
LDI	Landing Direction Indicator
LDIN	Lead In lighting system
LDN	Level Day Night
LE	Leading Edge
LED	Light Emitting Diode
LEMAC	Leading Edge Mean Aerodynamic Chord
LF	Low Frequency
LFA	Low-Frequency radio range

LFM	Low-powered Fan Marker
LFR	Low-Frequency Range
lgtd	lighted
lgth	length
lgts	lights
LH	Left Hand
LIM	Compass Locator (at ILS) Inner Marker
liq	liquid
LIRL	Low-Intensity Runway edge Lights
LL	Low Lead
LLWAS	Low-Level Wind shear Alert System
llz	localizer
LMM	Compass Locator (at ILS) Middle Marker
LMT	Local Mean Time
lndg	landing
lng	landing
LO	Locator Outer marker
LOA	Letter Of Authorization
loc	localizer, location, locator

LOC-BC	Localizer Back Course
LOFT	Line-Oriented Flight Training
LOID	Location Identifiers
LOM	Compass Locator (at ILS) Outer Marker
long	longitude
LOP	Line Of Position
LORAN	Long-Range Navigation
LOX	Liquid Oxygen
LPT	Low-Pressure Turbine
LR	Lapse Rate
LR	Lead Radial
LR	Long Range
LRA	Landing Rights Airport
LRCO	Limited Remote Communications Outlet
LRF	Long-Range Fuel
lrn	loran
LRNAV	Long-Range Navigation
LRR	Long-Range Radar
LRU	Line Replaceable Unit
LSALT	Lowest Safe Altitude

LSB Lower Side Band

LSS Local Speed of Sound

LT Left Turn after take-off

LT Local Time

lts lights

lube lubricate

LVAR Thrust Lever Angle Resolver

LVDT Linear Variable Differential Transformer

LVOR Low VOR

Lyc Lycoming (engine manufacturer)

M MACH

M Maintain

M Maintenance

M Meters

MAA Maximum Authorized Altitude

MAC Mean Aerodynamic Chord

MAC Michigan Aeronautics Commission

MAC Military Airlift Command

MAD Magnetic Anomaly Director

mag magnesium, magnetic, magneto

maint	maintain, maintenance
MALS	Medium intensity Approach Lighting System
MALSF	Medium intensity Approach Lighting System with sequenced Flashing lights
MALSR	Medium intensity Approach Lighting System with Runway alignment indicator lights
man	manual
map	mapping
MAP	Manifold Absolute Pressure
MAP	Missed Approach Point
Mar	March
MAR	Modernization and Association Restrictions
MARSA	Military Authority assumes Responsibility
matl	material
MATOGW	Maximum Allowable Take-Off Gross Weight
MB	Magnetic Bearing
MB	Marker Beacon
mb	millibars

MBOH	Minimum Break Off Height
MBR	Marker Beacon Receiver
MBZ	Mandatory Broadcast Zone
MC	Magnetic Course
MCA	Minimum Controllable Airspeed
MCA	Minimum Crossing Altitude
MCAF	Marine Corps Air Facility
MCAS	Marine Corps Air Station
MCTA	Military Controlled Airspace
MD	McDonnell Douglas
MDA	Minimum Descent Altitude
MDH	Minimum Descent Height
MDT	Mountain Daylight Time
ME	Mechanical Engineer
ME	Multi-Engine
MEA	Minimum En route Altitude
mech	mechanical
med	medium
MEHT	Minimum Eye Height over Threshold
MEI	Multi-Engine Instructor
MEL	Minimum Equipment List

MEL	Multi-Engine Land
mem	memorial
meml	memorial
MEP	Mean Effective Pressure
met	meteorological
METO	Maximum Except Take-Off (power)
MF	Mandatory Frequency
MF	Medium Frequency
mfd	manufactured
MFD	Multifunctional Display
MFP	Mass Flow Parameter
mfr	manufacturer
MFR	Medium Frequency Range
MFT	Meter Fix Time/slot time
mgr	manager
MGW	Maximum Gross Weight
MH	Magnetic Heading
MHA	Minimum Holding Altitude
MHz	Megahertz
MI	Indicated MACH number
MI	Medium Intensity
mi	mile

MIA	Minimum IFR Altitude
MIALS	Medium Intensity Approach Light System
MID/ASIA	Middle East/Asia region (ICAO)
mil	military
MIL SPEC	Military Specification
mim	minimum
min	minimum, minute
minima	minimums
MIRL	Medium Intensity Runway edge Lights
MIS	Meterological Impact Statement
mix	mixture
mkr	marker
MLDI	Meter List Display Interval
MLID	Monitor Location Identifier
MLS	Microwave Landing System
MLW	Maximum Landing Weight
MM	Middle Marker
mm	millimeters
MMEL	Master Minimum Equipment List
MN	Magnetic North

MNP	Magnetic North Pole
MNPS	Minimum Navigation Performance Specification
MNPSA	Minimum Navigation Performance Specification Airspace
MOA	Military Operations Area
MOCA	Minimum Obstruction Clearance Altitude
mod	modulate, modulation
mom	moment
MON	Motor Octane Number
MOPS	Minimum Operating Performance Standard
MORA	Minimum Off Route Altitude
mos	months
MP	Manifold Pressure
MPH	Miles Per Hour
MPP	Most Probable Position
MPS	Meters Per Second
MPU	Microprocessor Unit
MPW	Maximum Permitted Weight
MRA	Minimum Reception Altitude
MRB	Material Review Board

MS	Margin of Safety
MS	Material Standard
MS	Military Standard
MSA	Minimum Safe Altitude
MSA	Minimum Sector Altitude
MSAW	Minimum Safe Altitude Warning
MSDS	Material Safety Data Sheet
MSL	Mean Sea Level
MSP	Magnetic South Pole
MST	Mountain Standard Time
MT	Metric Ton
MTA	Military Training Area
MTCA	Minimum Terrain Clearance Altitude
mtg	mounting
MTI	Moving Target Indicator
MTOW	Maximum Take-Off Weight
MTR	Military Training Route
MTWA	Maximum Total Weight Authorized
MU	Friction value representing runway surface conditions (designation)
mun	municipal

muni	municipal
MUTA	Military Upper Traffic control Area
MVA	Minimum Vectoring Altitude
MWO	Meteorological Watch Office
MZFW	Maximum Zero Fuel Weight

N	Night
N	North
N/A	Not Applicable
NA	North America(n)
NA	Not Authorized
NA	Not Available
NAA	National Aeronautics Association
NAAS	Naval Auxiliary Air Station
nac	nacelle
NACA	National Advisory Committee for Aeronautics (now NASA)
NADC	Naval Air Development Center
NAEC	Naval Air Engineering Center
NAF	Naval Air Facility
NAFI	National Association of Flight Instructors

NALF	Naval Auxiliary Landing Field
NAM	Nautical Air Miles
NAM	North American region (ICAO)
NAP	Noise Abatement Procedure
NAR	North American Route
NAR	North Atlantic Route
NARL	Naval Arctic Research Laboratory
NAS	National Aircraft Standard
NAS	National Airspace System
NAS	Naval Air Station
NASA	National Aeronautics and Space Administration
NASAO	National Association of State Aviation—Officials
NASC	National Aircraft Standards Committee
NASP	National Aerospace Plane
NAT	North Atlantic Traffic
NATA	National Air Transportation Association
NATA	National Aviation Trades Association

NATCO	Northwest Aerospace Training Corporation
NATCOM	National Communication center
NATO	North American Treaty Organization
NATOTS	North Atlantic Organized Track System
natl	national
nav	navigation, navigator
NAVAID	Navigational Aid
NAWAU	National Aviation Weather Advisory Unit
NBAA	National Business Aircraft Association
NBCAP	National Beacon Code Allocation Plan
nbnd	northbound
NBS	National Bureau Standards
NCA	Northern Control Area
NCRP	Noncompulsory Reporting Point
NDB	Nondirectional Beacon
NDH	No Damage History

NDPER	National Designated Pilot Examiner Registry
NE	Northeast
neg	negative
NESDIS	National Environmental Satellite Data and Information Service
NEXRAD	Next generation weather Radar
NFCT	Nonfederal Control Tower
NFDC	National Flight Data Center
NFDD	National Flight Data Digest
NHC	National Hurricane Center
NIFA	National Intercollegiate Flying Association
NLP	Normal Lead Point
NM	Nautical Mile
NMAC	Near Mid-Air Collision
NMC	National Meteorological Center
NMI	Nautical Mile
nml	normal
NMPH	Nautical Miles Per Hour
NMR	Nautical Mile Radius
NMS	Navigational Management System

NNE	North-Northeast
no	number
NOAA	National Oceanic and Atmospheric Administration
NOMAD	Navy Oceanographic and Meteorological Device
NOPAC	North Pacific
NoPT	No Procedure Turn
NORDO	No Radio
NOS	National Ocean Service
NOTAM	Notices To Airman
Nov	November
NPRM	Notices of Proposed Rule Making
NPS	National Park Service
NSSFC	National Severe Storms Forecast Center
nstd	nonstandard
NTAP	Notices To Airman Publication
NTC	Negative Torque Control
ntc	notice
NTIS	National Technical Information Service
NTS	Negative Torque Sensing system

NTSB	National Transportation Safety Board
NTZ	No Transgression Zone
NW	Northwest
NWC	Naval Weapons Center
NWS	National Weather Service

O	Operations
O/A	On or About
O/R	On Request
O/T	Other Times
OALT	Operational Acceptance Level of Traffic
OAM	Office of Aviation Medicine
OAP	Offset Aim Point
OAT	Outside Air Temperature
OBS	Omnibearing Selector
obst	obstacle, obstruction
obstn	obstruction
OC	Overcentering
OCA	Obstacle Clearance Altitude
OCH	Obstacle Clearance Height

OCL	Obstacle Clearance Limit
OCS	Officer Candidate School
Oct	October
OD	Outside Diameter
ODALS	Omnidirectional Approach Lighting Systems
ODAPS	Oceanic Display And Planning System
ODS	Operator-input and Display System
OEI	One Engine Inoperative
ofc	office
OFT	Outer Fix Time
OFZ	Obstacle Free Zone
OGE	Out of Ground Effect
OGV	Outlet Guide Vanes
OK	Operating normally
OLDI	On Line Data Interchange
OLF	Outlying Field
OLS	Optical Landing System
OM	Outer Marker
ON	Octane Number
ONC	Operational Navigational Chart

ONER	Oceanic Navigational Error Report
op	operation
oper	operate, operating
OPI	Office of Primary Interest
opn	operation
opns	operations
opp	opposite
opr	operate
ops	operates, operation
opt	option(al)
orig	original
OSHA	Occupational Safety and Health Administration
OST	Office of the Secretary of Transportation
OTC	Over The Counter
OTR	Oceanic Transition Route
OTS	Organized Track System
OTS	Out of Service
outbd	outboard
ovrn	overrun
oxy	oxygen

P	Phase
P	Piston
P	Pitot
P	Precision
P	Prohibited
P	Proposed
P&W	Pratt & Whitney (engine manufacturer)
P-Line	Pole Line
P-Lns	Power Lines
P/T	Power Turbine
PA	Pressure Altitude
PAC	Pacific region (ICAO)
PACE	Pilot and Aircraft Courtesy Evaluation
PAEW	Personnel And Equipment Working
PAJA	Parachute Jumping Activities
PAMA	Professional Aviation Maintenance Association
PAPI	Precision Approach Path Indicator
par	parallel
PAR	Precision Approach Radar

PAR	Preferential Arrival Route
parl	parallel
pass	passenger
pat	pattern
patt	pattern
PATWAS	Pilots Automatic Telephone Weather Answering Service
pax	passenger
PB	Pressure Breathing
PBCT	Proposed Boundary Crossing Time
PBPF	Propeller Blade Passing Frequency
PCA	Polar Cap Anomaly
PCA	Positive Control Area
PCL	Pilot Controlled Lighting
PCN	Pavement Classification Number
PCU	Propeller Control Unit
PCZ	Positive Control Zone
PDAR	Preferential Departure and Arrival Route
PDC	Predeparture Clearance
PDIU	Propulsion Discrete Interface Unit
PDR	Preferential Departure Route

PDT	Pacific Daylight Time
PE	Potential Energy
PE	Power Enrichment
PEP	Piper External Power
perm	permanent(ly)
PEST	Pilot Examination Standardization Team
PFC	Porous Friction Courses
PFD	Primary Flight Display
PFE	Professional Flight Engineer
PGC	Propeller Gear Case
PHAK	Pilot's Handbook of Aeronautical Knowledge
PIC	Pilot In Command
PIDP	Programmable Indicator Data Processor
PIREP	Pilot Report
PJE	Parachute Jumping Exercise
PL	Parts List
PL	Power Lever
PLA	Power Lever Angle
PLA	Practice Low Approach

PLASI	Pulsating visual Approach Slope Indicator
PMA	Parts Manufacturer Approval
PMA	Permanent Magnet Alternator
PMMEL	Proposed Master Minimum Equipment List
PN	Prior Notice required
PNI	Pictorial Navigation Indicator
PNR	Point of No Return
POH	Pilots Operating Handbook
POM	Pilots Operating Manual
pos	position, positive
posn	position
PPH	Pounds Per Hour
PPI	Plan Position Indicator
PPO	Prior Permission Only
PPR	Prior Permission Required
PPS	Pulses Per Second
PR	Pressure Ratio
PRA	Precision Radar Approach
press	pressure
prev	previous

PRF	Pulse Repetition Frequency
PRIRA	Primary Radar
PRM	Precision Runway Monitor
proc	procedure
proj	projection
prop	propeller
prov	provisional
PSF	Pounds per Square Foot
psgr	passenger
PSI	Pounds per Square Inch
PSIA	Pounds per Square Inch Absolute
PSID	Differential Pressure
PSR	Packed Snow on Runway
PST	Pacific Standard Time
PT	Procedure Turn
PTB	Pounds per Thousand Barrels
PTC	Pretaxi Clearance
PTN	Procedure Turn
PTO	Part-Time Operation
PTO	Power Take-Off assembly
PTS	Polar Track Structure

PTS	Practical Test Standards
PTT	Push To Talk
PVASI	Pulsating Visual Approach Slope Indicator
pvt	private
pwr	power
pyro	pyrotechnics
QDM	Magnetic great circle bearing of the station from the aircraft
QDR	Magnetic great circle bearing of the aircraft from the station
QFE	Airfield datum pressure
QFF	Mean sea level pressure
QNE	Barometric pressure standard
QNH	Barometric pressure at a particular station
QTE	True great circle bearing of the aircraft from the station
qty	quantity
QUJ	True great circle bearing of the station from the aircraft

R	Radar
R	Radial
R	Radius
R	Restricted
R	Right
R12	Freon for air conditioning (old style)
R134A	Environmentally safe air conditioning freon
RA	Radio Altimeter
RA	Resolution Advisory
RAA	Regional Airline Association
RAAF	Royal Australian Air Force
RADAR	Radio Detection And Ranging
RAF	Royal Air Force
RAFC	Regional Area Forecast Center
RAI	Runway Alignment Indicator
RAIL	Runway Alignment Indicator Lights
RAMOS	Remote Automatic Meteorological Observation System
RAPCON	Radar Approach Control
RAR	Radar Arrival Route

RAS	Rectified Airspeed
RAS	Runway Alert System
RAT	Ram Air Temperature
RATCF	Radar Air Traffic Control Facility
RB	Relative Bearing
RBN	Radio Beacon
RC	Radio Controlled (usually model aircraft)
RC	Rate of Climb
RCAG	Remote Communications Air/Ground facility
RCC	Rescue Coordination Center
RCDI	Rate of Climb and Descent Indicator
RCL	Runway Centerline
RCLM	Runway Centerline Marking
RCLS	Runway Centerline Light System
RCO	Remote Communication Outlet
RCR	Reverse Current Relay
RCR	Runway Condition Reading
rcv	receive
rcvr	receiver
rcvs	receives

RD	Real Drift
R&D	Research and Development
rdo	radio
RDR	Radar Departure Route
recept	receptacle
recip	reciprocating
rect	rectangular
ref	reference
reg	regular, regulation, regulatory
REIL	Runway End Identifier Lights
reinf	reinforced
rel	release
relctd	relocated
rem	remaining
REP	Reporting Point
req	request, required
RETA	Revised Estimated Time of Arrival
retr	retractable
rev	revolution
RF	Radio Facility
RFC	Royal Flying Corps
rflg	refueling

RFM	Rotorcraft Flight Manual
RFSC	Rubberized Friction Seal Coat
RG	Retractable Gear
rge	range
RGT TRFC	Right Traffic
RH	Right Hand
RIC	Remote Indicating Compass
RIS	Report Identification Symbol
RL	Low-intensity Runway Lights or intensity not specified
RLCE	Request Level Change En route
RLG	Ring Laser Gyro
RLLS	Runway Lead-in Lighting System
rls	release
rm	remarks
RMI	Radio Magnetic Indicator
rmk	remark
RMS	Root-Mean-Square
rmvl	removal
RNAV	Random and fixed Area Navigation

RNPC	Required Navigation Performance Capability
rntl	rental
ROC	Rate Of Climb
RON	Remaining Overnight
RON	Research Octane Number
ROM	Read Only Memory
ROTC	Reserve Officers Training Corps
rotg	rotating
RPI	Runway Point of Intercept(ion)
RPK	Revenue Passenger Kilometer
RPM	Revenue Passenger Mile
RPM	Revolutions Per Minute
rpr	repairs
rprt	report
RPS	Revolutions Per Second
RPU	Receiver/Processor Unit
rqr	require
rqrd	required
RR	low or medium frequency Radio Range station
RR	Railroad

RRL	Runway Remaining Lighting
RRWDS	Radar Remote Weather Display System
RRZ	Radar Regulation Zone
RSA	Runway Safety Area
RSC	Runway Surface Condition
RSP	Responder Beacon
RSPA	Research and Special Programs Administration
rsvn	reservation
RT	Radar Tracking
RT	Right Turn after take-off
RTCA	Radio Technical Commission for Aeronautics
rte	route
rtf	radiotelephone
rtg	radiotelegraph
RTR	Remote Transmitter/Receiver
RTS	Return To Service
rubber accum	rubber accumulation
rud	rudder
RV	Radar Vector

RVP Reid Vapor Pressure
RVR Runway Visual Range
RVRM Runway Visual Range Midpoint
RVRR Runway Visual Range Rollout
RVRT Runway Visual Range Touchdown
RVV Runway Visual Value
RW Random Wander
RW Relative Wind
rwy runway
RxR Railroad
ry runway

S Single-wheel (landing gear)
S South
S Straight-in
S Surveyed
SA Selective Availability
SAE Society of Automotive Engineering
SALR Saturated Adiabatic Lapse Rate
SALS Short Approach Light System
SALSF Short Approach Lighting System with sequenced Flashing lights

SAM	South American region (ICAO)
SAO	Surface Aviation Observation
SAP	Soon As Possible
SAR	Search And Rescue
SARDA	State And Regional Disaster Airlift
SASE	Self-Addressed Stamped Envelope
SAT	Static Air Temperature
SAUS	Surface Aviation observation United States
SAVASI	Simplified Abbreviated Visual Approach Slope Indicator
SAWRS	Supplemental Aviation Weather Reporting Station
SB	Service Bulletin
sbnd	Southbound
sby	standby
SCATANA	Security Control of Air Traffic and Air Navigation Aids
SCMOH	Since Chrome Major Overhaul
SCOH	Since Chrome Overhaul

SCR	Special Certification Review
SCSG	Silicon-Chip Strain Guage
SCTOH	Since Chrome Top Overhaul
SCUBA	Self-Contained Underwater Breathing Apparatus
SCWD	Super-Cooled Water Droplet
SDAC	Safety and Data Analysis unit
SDF	Simplified Directional Facility
SDR	Service Difficulty Report
SE	Single Engine
SE	Southeast
sec	second
SECRA	Secondary Radar
sect	section
SEIC	System Engineering and Integration Contract
sel	selector
SELCAL	Selective Calling
sep	separate
Sep	September
seq	sequence
ser	service

SER	Stop End of Runway
SFA	Single-Frequency Approach
SFAR	Special Federal Aviation Regulation
SFC	Specific Fuel Consumption
sfc	surface
SFDF	Satellite Field Distribution Facility
SFL	Sequenced Flashing Lights
SFO	Simulated Flameout
SFO	Single-Frequency Outlet
SFRMN	Since Factory Remanufactured
SFSS	Satellite Field Service Station
SG	Specific Gravity
SHF	Super-High Frequency
SHP	Shaft Horsepower
SI	Spark Ignited
SI	Straight-In
SIAP	Standard Instrument Approach Procedure
SID	Standard Instrument Departure
SID	Sudden Ionospheric Disturbance
SIT	Self-Ignition Temperature

SIWL	Single Isolated Wheel Load
sk	sky
skd	scheduled
SL	Sea Level
SLDI	Sector List Drop Interval
SLS	Side Lobe Suppression
SM	Statute Mile
SMGCS	Surface Movement Guidance and Control System
SMOH	Since Major Overhaul
SN	Strategic Navigation
snd	sand, sanded
SNEW	Since New
sngl	single
SNR	Signal-to-Noise Ratio
SNW RMVL	Snow Removal
SOC	Start Of Climb
SODALS	Simplified Omnidirectional Approach Lighting System
SOH	Since Overhaul
SOIR	Simultaneous Operations on Intersecting Runways

spd	speed
spec	specification
SPOH	Since Prop Overhaul
sprvsr	supervisor
sq	square, squawk
sqdn	squadron
SR	Short Range
SR	Specific Range
sr	sunrise
SRA	Special Rules Area
SRA	Surveillance Radar Approach
SRE	Surveillance Radar Element
SRS	Same Runway Separation
SS	Steady State
ss	sunset
SSALF	Simplified Short Approach Light system with sequenced Flashing lights
SSALR	Simplified Short Approach Light system with Runway alignment indicator lights
SSALS	Simplified Short Approach Light System

SSB	Single-Sided Band
SSE	South-Southeast
SSR	Secondary Surveillance Radar
SST	Supersonic Transport
SSV	Standard Service Volume
ST	Static Thrust
sta	station
stab	stabilizer
STAR	Standard Terminal Arrival Route
stby	standby
STC	Sensitivity Time Control
STC	Supplemental Type Certificate
std	standard
STOH	Since Top Overhaul
STOL	Short Take-Off and Landing
STOP	Since Top overhaul
sts	status
stud	student
SUPPS	Supplemental Procedures
surf	surface
SURPIC	Surface Picture
SVA	Stator Vane Actuators

svc	service
SVFR	Special VFR
SW	Single-Wheel landing gear
SW	Southwest
sw	switch
SWAP	Severe Weather Avoidance Plan
SWSL	Supplemental Weather Service Location
swy	stopway
sync	synchronize
syst	system

T	Terminal
T	Transmits only
T	Turboprop
T&B	Turn and Bank
T-VASI	Tee Visual Approach Slope Indicator
T/F	Turbofan
T/J	Turbojet
T/O	Take-Off
T/S	Turboshaft

T/SI	Turn and Slip Indicator
T/W	Tailwind
TA	Temperature
TA	Traffic Advisory
TA	Transition Altitude
TA	True Altitude
TAAS	Terminal Advanced Automation System
TAC	TACAN
TAC	Tactical Air Command
TAC	Terminal Area Chart
TACAN	Tactical Air Navigation (UHF)
tach	tachometer
TAP	Total Air Pressure
TAR	Terminal Area surveillance Radar
TAS	True Airspeed
TAT	Total Air Temperature
tb	turbulence
TBI	Turn and Bank Indicator
TBO	Time Between Overhauls
TC	Top Center
TC	True Course

TC	Turbocharger
TC	Type Certificate
TCA	Terminal Control Area
TCAS	Terminal Collision Avoidance System
TCAS	Traffic alert and Collision Avoidance System
TCDS	Type Certificate Data Sheet
TCH	Threshold Crossing Height
TCLT	Tentative Calculated Landing Time
TCTA	Transcontinental Control Area
TD	Technical Data
TD	Time Difference
TD	Total Drift
TDC	Top Dead Center
TDWR	Terminal Doppler Weather Radar
TDZ	Touchdown Zone
TDZ/CL	Touchdown Zone and runway Centerline Lighting
TDZE	Touchdown Zone Elevation
TDZL	Touchdown Zone Lights
TE	Total Energy

TE	Trailing Edge
TEA	Track Error Angle
TEC	Tower En route Control
tel	telephone
TEMAC	Trailing Edge Mean Aerodynamic Chord
temp	temporary
TERPS	Terminal instrument Procedures
TES	Thermal Energy Storage
tet	tetrahedron
TET	Turbine Exhaust Temperature
TF	Terrain Following
tfc	traffic
TFR	Temporary Flight Restriction
TGL	Touch and Go Landings
TGT	Turbine Gas Temperature
TH	True Heading
thld	threshold
THP	Thrust Horsepower
thr	threshold
TIBS	Transcribed Information Briefing Service

TIP	Turbine Inlet Pressure
TIS	Time In Service
TIT	Turbine Inlet Temperature
tk	track
TKE	Track Error Angle
tkf	take-off
tkof	take-off
TL	Thrust Line
TL	Transition Level
TLv	Transition Level
TM	Technical Memorandum
tm	time
TMA	Terminal control Area
TMC	Terminal control Center
tml	terminal
TMPA	Traffic Management Program Alert
tmpry	temporarily, temporary
TMU	Traffic Management Unit
TN	Technical Note
TN	True North
TO	Take-Off
TO	Technical Order

TODA	Take-Off Distance Available
TOP	Turbine Outlet Pressure
TORA	Take-Off Run Available
TOS	Traffic Orientation System
TOT	Turbine Outlet Temperature
tp	type
TPA	Traffic Pattern Altitude
TPP	U.S. government Terminal Procedures Publication
tr	track
TR	Technical Report
TR	Transmit-Receive
TRA	Temporary Reserved Airspace
TRACON	Terminal Radar Approach Control
trans	transition
transp	transportation
TRCV	Tri-Color Visual approach slope indicator
trng	training
trnsp	transportation
TRSA	Terminal Radar Service Area

TRSB	Time Reference Scanning Beam
trsn	transition
trtd	treated
TS	Tensile Strength
TSFC	Thrust Specific Fuel Consumption
tsnt	transient
TSO	Technical Standards Order
tt	teletypewriter
TT	Total Time
TTAE	Total Time Aircraft Engine
TTAF	Total Time Aircraft Frame
TTG	Time To Go
TTS	Time To Station
TTSN	Total Time Since New
TUC	Time of Useful Consciousness
TVCS	Thrust Vectoring Control System
TVOR	Terminal Very high frequency Omnidirectional Range station
TW	Transport Wander
TWEB	Transcribed Weather En route Briefing
twp	township

twr	tower
TWU	Transport Workers Union
twy	taxiway
txp	transponder
txpdr	transponder

U	Unicom
U/S	Unserviceable
UAA	University Aviation Association
UAA	Upper Advisory Area
UACC	Upper Area Control Center
UAR	Upper Air Route
UDF	(UHF) Direction Finder
UDF	Unducted Fan
UDP	Upper Deck Pressure
UFN	Until Further Notice
UFO	Unidentified Flying Object
UHB	Ultra-High Bypass
UHF	Ultra-High Frequency
UK	United Kingdom
ukn	unknown

ult	ultimate
unatnd	unattended
unavbl	unavailable
UNICOM	Aeronautical advisory station
unkn	unknown
unlgtd	unlighted
unmkd	unmarked
unmon	unmonitored
unrelbl	unreliable
unusbl	unusable
US	United States
USA	United States of America
USAAF	United States Army Air Force
USAATS	United States of America Air Traffic Service
USAF	United States Air Force
USAFIB	(USA) Army-Aviation Flight Information Bulletin
USAFR	United States Air Force Reserve
USATS	United States Air Traffic Service
USB	Upper Side Band
USMC	United States Marine Corps

USN	United States Navy
USNA	United States National Army
USNR	United States Navy Reserve
USO	United Service Organization
UTC	Coordinated Universal Time
UUA	Urgent pilot report (message identifier)
UWY	Upper airway
V	Velocity
V-Speeds	Performance speeds
V/V	Vertical Velocity
vac	vacuum
VAFTAD	Volcanic Ash Forecast Transport And Dispersion
VAPC	Variable Absolute Pressure Controller
VAPI	Visual Approach Path Indicator
var	variation
VASI	Visual Approach Slope Indicator
VCL	Vertical Component of Lift
vcnty	vicinity

VCW	Vertical Component of Weight
VDF	(VHF) Direction Finder
VDP	Visual Descent Point
vel	velocity
vert	vertical
VFR	Visual Flight Rules
VFW	Veteran(s) of Foreign Wars
vfy	verify
VG	Variable Geometry
VG	Velocity vs. Gravity
VGV	Variable Guide Vanes
VHF	Very High Frequency
VI	Viscosity Index
via	by way of
vice	instead, versus
VIP	Video Integrator Processor
VLF	Very Low Frequency
VMC	Visual Meteorological Conditions
VNAP	Vertical Noise Abatement Procedures
VNAV	Vertical Navigation
vol	volume

VOLMET	Meteorological information for aircraft in flight
VOR	(VHF) Omnidirectional Radio range
VOR/DME	Collocated VOR and DME NAVAIDS
VORTAC	Collocated VOR and TACAN
VORW	VOR (Without voice)
VOT	(VHF) Omnidirectional Test (VOR test signal)
VR	(VFR) military training Route
VR	Visual Reference
VR	Voltage Regulator
VRA	Vertical Resolution Advisory
VRS	Voice Response System
VSCS	Voice Switching and Control System
VSI	Vertical Speed Indicator
VSV	Variable Stator Vanes
Vt	Terminal Velocity
VTA	Vertex Time of Arrival
VTOL	Vertical Take-Off and Landing
VVI	Vertical Velocity Indicator

W	Warning
W	Weight
W	West
W	Without voice
W&B	Weight and Balance
W/B	Weight and Balance
W/D	Weight/Drag ratio
W/O	Without
W/P	Waypoint
W/V	Wind Vector
WA	Weather Advisory (AIRMET message identifier)
WAC	World Aeronautical Charts
WAFS	World Area Forecast System
WATS	Wide Area Telephone Service
wbnd	Westbound
WC	Wire Combed
WCA	Wind Correction Angle
WD	Wind Direction
WDI	Wind Direction Indicator
we	weekend
WEF	With Effect From

WFMU	Weather Fixed Map Unit
WFO	Weather Forecast Office
WGS	World Geodetic System
wgt	weight
wi	within
WIE	With Immediate Effects
WILCO	Will Comply
WIP	Work In Progress
wkdays	weekdays (Monday through Friday)
wkends	weekends (Saturday and Sunday)
WMO	World Meteorological Organization
WMSC	Weather Message Switching Center
WMU	Western Michigan University (School of Aviation Science)
WOT	Wide Open Throttle
wp	waypoint
wpt	waypoint
wrn	Western
WSFO	Weather Service Forecast Office
WSO	Weather Service Office
WSR	Weather Service Radar

WST	Convective SIGMET (message identifier)
wt	weight
WTM	Wind Tunnel Model
WV	Wind Velocity
WVC	Water Vapor Content
WWI	World War I
WWII	World War II

X	Cross
X	On request
X	Radar only frequency
X/W	Crosswind
xcpt	except
xmits	transmits
xpct	expect
xpdr	transponder
xpnd	transponder
xpndr	transponder
XTK	Cross-Track

yds yards
YP Yield Point

Z ZULU designator for (UTC)
ZFW Zero Fuel Weight

TWO

Weather-Related Abbreviations and Symbols

WX SYMBOLS

"WEA"	Indicates manual observer data
$	Maintenance check indicator
(H)	Highest
(L)	Lowest
(M)	Middle
+	Heavy, Increasing
++	Very heavy
-	Light, Decreasing
- -	Very light
-BKN	Thin Broken
-OVC	Thin Overcast
-SCT	Thin Scattered
-X	Surface based obstruction is partially Obscured
-	Through
( )	No symbol=Moderate
X	Intense
XX	Extreme
0	Positive
1	Negative

ABBREVIATIONS

A	Absolute
A	Aircraft-reported ceiling
A	Alaskan Standard Time
A	Altimeter setting in inches
A	Arctic
A	Hail
AAWF	Auxillary Aviation Weather Facility
ABT	About
ABV	Above
AC	Altocumulus
AC	Convective Outlook Bulletin
ACC	Altocumulus Castellanus
ACCAS	Altocumulus Castellanus
ACLD	Above Clouds
ACLS	Standing Lenticular Altocumulus
ACPY	Accompany
ACRS	Across
ACSL	Altocumulus Standing Lenticular
ACT	Active
ACTG	Acting
ACTV	Active

Weather-Related Abbreviations and Symbols

ACYC	Anticyclonic
ADV	Advise
ADVCTN	Advection
ADVN	Advance
ADVY	Advisory
ADZ	Advise
ADZD	Advised
ADZY	Advisory
AFC	Area Forecast Center
AFCT	Affect
AFDK	After Dark
AFT	After
AFTN	Afternoon
AGN	Again
AHD	Ahead
AK	Alaska
AL	Alabama
ALF	Aloft
ALG	Along
ALGHNY	Allegheny
ALQDS	All Quadrants
ALREPS	Airline Reports

ALSEC	All Sectors
ALSTG	Altimeter Setting
ALT	Altitude
ALTA	Alberta
ALTN	Alternate
ALTNLY	Alternately
ALUTN	Aleutian
ALWF	Actual Wind Factor
AM	Ante Meridiem
Am	Arctic Maritime
AMD	Amended, Amendment
AMPLTD	Amplitude
AMS	Airmass
AMS	American Meteorological Society
AMT	Amount
ANLYS	Analysis
AO	Designation for AWOS
AO2	Unattended (no observer) ASOS
AO2A	Attended (observer present) ASOS
AOA	At Or Above
AOB	At Or Below

AP	Small Hail
APCH	Approach
APCHG	Approaching
APLCN	Appalachian
AR	Arkansas
ARND	Around
ARNO	Azimuth/Range indicator not operative
AS	Altostratus
ASSW	Associated With
ATLC	Atlantic
AURBO	Aurora Borealis
AUTO	Automated Observation
AVG	Average
AWP	Aviation Weather Processors
AWW	Alert message
AWY	Airway
AZ	Arizona
B	Balloon reported ceiling
B	Began (time weather started, in minutes)

B	Bering Standard Time
BACLIN	Baroclinic
BATROP	Barotropic
BC	British Columbia
BC	Patches
BCFG	Fog patches
BCH	Beach
BCKG	Backing
BCM	Become
BCMG	Becoming
BD	Blowing Dust
BDA	Bermuda
BDR	Border
BECMG	Becoming
BFDK	Before Dark
BFR	Before
BGN	Began, Begin
BHND	Behind
BINOVC	Breaks In Overcast
BKN	Broken ($5/8$ to $7/8$ cloud coverage)
BL	Between Layers
BL	Blowing

BLD	Build
BLDUP	Buildup
BLKHLS	Black Hills
BLKT	Blanket
BLO	Below
BLW	Below
BLZD	Blizzard
BMS	Basic Meteorological Services
BN	Blowing sand
BNDRY	Boundary
BNK	Bank
BNTH	Beneath
BOVC	Base of Overcast
BR	Mist (visibility (0.625–6 mi)
BRAF	Braking Action Fair
BRAG	Braking Action Good
BRAN	Braking Action Nil
BRAP	Braking Action Poor
BRF	Brief
BRK	Break
BRKHIC	Breaks in Higher Overcast
BRKN	Broken

BRKS	Breaks
BRKSHR	Berkshires
BRM	Barometer
BS	Blowing Snow
BTN	Between
BTR	Better
BTW	Between
BTWN	Between
BWER	Bounded Weak Echo Region
BY	Blowing spray
BYD	Beyond

C	Celsius
C	Center
C	Central
C	Central Standard Time
C	Cold
C	Continental
CA	California
CA	Continental Arctic
CAN	Canada

CARIB	Caribbean
CASCDS	Cascades
CAT	Clear Air Turbulence
CAUFN	Caution
CAVOK	Cloud And Visibility are OK
CAVU	Ceiling And Visibility Unlimited
CAWS	Common Aviation Weather Subsystem
CB	Cumulonimbus
CBMAM	Cumulonimbus Mamma
CC	Cirrocumulus
CC	Cloud to Cloud
CCLKWS	Counterclockwise
CCSL	Standing Lenticular Cirrocumulus
CDFNT	Cold Front
CFN	Confine
CFP	Cold Front Passage
CG	Cloud to Ground
CHARC	Characteristics
CHC	Chance
CHG	Change
CHSPK	Chesapeake

CI	Cirrus
CIG	Ceiling
CLD	Cloud
CLDS	Clouds
CLR	Clear ($^0/_8$ cloud coverage)
CLRG	Clearing
CLRS	Clear and Smooth
CM	Cumulonimbus Mamma
CMNC	Commence
CNCL	Cancel
CND	Condition
CNDN	Canadian
CNS	Continuous
CNTR	Center
CNTRL	Central
CNVTV	Convective
CO	Colorado
CON	Condensation
COND	Condition
CONFDC	Confidence
CONT	Continue, Continuous
CONTDVD	Continental Divide

CONTG	Continuing
CONTLY	Continuously
CONTRAILS	Condensation Trails
COR	Correction
COT	At or on the Coast
COT	Continuous
CRLCN	Circulation
CS	Cirrostratus
CSDRBL	Considerable
CST	Coast
CSTL	Coastal
CT	Connecticut
CTGY	Category
CTSKLS	Catskills
CU	Cumulus
CUFRA	Cumulus Fractus
CVR	Cover
CYC	Cyclonic
CYCLGN	Cyclogenesis
D	Down
D	Dust

D	Estimated cirriform ceiling
DABRK	Daybreak
DALGT	Daylight
DC	District of Columbia
DCAVU	Ceiling And Visibility Unlimited (with the remainder of the report missing)
DCR	Decrease
DCRG	Decreasing
DE	Delaware
DEG	Degree
DEP	Deep, Depth
DFUS	Diffuse
DIST	District
DKTS	Dakotas
DLA	Delay
DMSH	Diminish
DNS	Dense
DNSLP	Downslope
DNSTRM	Downstream
DO	Ditto
DP	Deep

DPN	Deepen
DPNG	Deepening
DPTH	Depth
DR	Low Drifting
DRFT	Drift
DRZL	Drizzle
DS	Dust Storm
DSIPT	Dissipate
DSIPTG	Dissipating
DSNT	Distant
DTLN	Dateline
DTRT	Deteriorate
DU	Dust
DURC	During Climb
DURD	During Descent
DURG	During
DVD	Divide
DVLP	Develop
DVV	Downward Vertical Velocity
DWNDFTS	Downdrafts
DWPNT	Dew Point
DZ	Drizzle

E	Eastern
E	Eastern Standard Time
E	Ended
E	Equatorial
E	Estimated
E	Sleet
ELNGT	Elongate
ELSW	Elsewhere
EMBD	Embedded
EMBDD	Embedded
EMSU	Environmental Meteorological Support Unit
ENDG	Ending
ENERN	East-Northeastern
ENEWD	East-Northeastward
ENG	England
ENTR	Entire
EOF	Expected Operations Forecast
ERY	Early
ESERN	Eastern
ESEWD	East-Southeastward
EST	Estimate

EVE	Evening
EW	Sleet showers
EWD	Eastward
EXCP	Except
EXPC	Expect
EXTRAP	Extrapolate
EXTRM	Extreme
EXTSV	Extensive

F	Fahrenheit
F	Fog
+F	Tornado/Waterspout
FA	Area Forecast
FAH	Fahrenheit
FC	Funnel Cloud
FCST	Forecast
FD	Winds and temperatures aloft forecast
FEW	$\frac{1}{8}$ or $\frac{2}{8}$ cloud coverage
FG	Fog
FIBI	Filed But Impracticable to transmit

FILG	Filing
FINO	Weather report will not be filed for transmission
FL	Flash advisory
FL	Florida
FLD	Field
FLDST	Flood Stage
FLG	Falling
FLRY	Flurry
FLW	Follow
FLWIS	Flood Warning Issued
FM	From
FNLN	Fine Line
FNT	Front
FNTGNS	Frontogenesis
FNTLYS	Frontolysis
FORNN	Forenoon
FPM	Feet Per Minute
FQT	Frequent
FR	Falling Rapidly
FRMG	Forming
FROPA	Frontal Passage

FROSFC	Frontal Surface
FRQ	Frequent
FRST	Frost
FRWF	Forecast Wind Factor
FRZ	Freeze
FRZLVL	Freezing Level
FRZN	Frozen
FT	Terminal Forecast
FTHR	Farther, Further
FU	Smoke
FWC	Fleet Weather Central
FWD	Forward
FZ	Freezing
G	Gust
GA	Georgia
GEN	General
GENLY	Generally
GF	Ground Fog
GFDEP	Ground Fog estimated Deep
GICG	Glaze Icing
GLFALSK	Gulf of Alaska

GLFCAL	Gulf of California
GLFMEX	Gulf of Mexico
GLFSTLAWR	Gulf of St. Lawrence
GND	Ground
GNDFG	Ground Fog
GR	Hail (0.25 in)
GRAD	Gradient
GRBNKS	Grand Banks
GRDL	Gradual
GRN	Ground
GRP	Group
GRTLKS	Great Lakes
GS	Small hail or snow pellets (0.25 in)
GSTS	Gusts
GSTY	Gusty
GV	Ground Visibility
H	Haze
H	Highest
HBRKN	High Broken

HCVIS	High Clouds Visible
HDEP	Haze layer estimated Deep
HDFRZ	Hard Freeze
HDSVLY	Hudson Valley
HGT	Height
HI	Hawaii
HI	High
HIEAT	Highest temperature Equaled for All Time
HIEFM	Highest temperature Equaled For the Month
HIER	Higher
HIESE	Highest temperature Equaled So Early
HIESL	Highest temperature Equaled So Late
HIFOR	High-level Forecast
HIR	Higher
HITMP	Highest Temperature
HIXAT	Highest temperature exceeded for All Time

HIXFM	Highest temperature exceeded For the Month
HIXSE	Highest temperature exceeded So Early
HIXSL	Highest temperature exceeded So Late
HLF	Half
HLSTO	Hailstones
HLTP	Hilltop
HLYR	Haze Layer aloft
HND	Hundred
HOVC	High Overcast
HRZN	Horizon
HSCTD	High Scattered
HURCN	Hurricane
HUREP	Hurricane Report
HVR	Heavier
HVY	Heavy
HVYR	Heavier
HX	Haze index
HYR	Higher
HZ	Haze

IA	Iowa
IC	Ice Crystals
ICE	Icing
ICG	Icing
ICGIC	Icing In Clouds
ICGICIP	Icing In Clouds In Precipitation
ICGIP	Icing In Precipitation
ID	Idaho
IF	Ice Fog
IL	Illinois
IMDT	Immediate
IMPT	Important
IN	Indiana
INC	In Cloud
INCR	Increase
INDC	Indicate
INDEF	Indefinite
INDEFLY	Indefinitely
INDFT	Indefinite
INFO	Information
INLD	Inland
INOPV	Inoperative

INSTBY	Instability
INTMT	Intermittent
INTR	Interior
INTRMTRGN	Inter-Mountain Region
INTS	Intense
INTSF	Intensifying
INTSFY	Intensify
INTST	Intensity
INVRN	Inversion
IOVC	In Overcast
IP	Ice Pellet
IPV	Improve
IPW	Ice Pellet showers
IR	Ice on Runway
IR	Infra-Red
ISOL	Isolated
ISOLD	Isolated
ITCZ	Inter-Tropical Convergence Zone
ITF	Inter-Tropical Front

JMP	Jump
JTSTR	Jet Stream
K	Cold
K	Prefix for contiguous United States
K	Smoke
KDEP	Smoke layer estimated Deep
KFRST	Killing Frost
KH	Smoke and Haze
KLYR	Smoke Layer aloft
KOCTY	Smoke Over City
KS	Kansas
KY	Kentucky
L	Drizzle
L	Left
L	Lowest
LA	Louisiana
LAN	Inland, Overland

LCL	Local
LCLLY	Locally
LCLY	Locally
LCTMP	Little Change in Temperature
LEWP	Line Echo Wave Pattern
LFT	Lift
LGRNG	Long Range
LGT	Light
LIFR	Low IFR
LK	Lake
LKLY	Likely
LLWS	Low-Level Wind Shear
LMT	Limit
LN	Line
LNS	Lines
LOC	Locally
LOEAT	Lowest temperature Equaled for All Time
LOEFM	Lowest temperature Equaled For the Month
LOESE	Lowest Temperature Equaled So Early

LOESL	Lowest temperature Equaled So Late
LOTMP	Lowest Temperature
LOXAT	Lowest temperature exceeded for All Time
LOXFM	Lowest temperature exceeded For the Month
LOXSE	Lowest temperature exceeded So Early
LOXSL	Lowest temperature exceeded So Late
LSQ	Line Squall
LSR	Loose Snow on Runway
LTG	Lightning
LTGCC	Lightning Cloud to Cloud
LTGCG	Lightning Cloud to Ground
LTGCW	Lightning Cloud to Water
LTGIC	Lightning In Clouds
LTGICCCCG	Lightning In Clouds, Cloud to Cloud, and Cloud to Ground
LTL	Little
LTLCG	Little Change
LTNG	Lightning

LTR	Later
LVL	Level
LWR	Lower
LWRBRKN	Lower Broken
LWROVC	Lower Overcast
LX	Low index
LYR	Layer, Layered
LYRD	Layered
M	Maritime
M	Measured
M	Middle
M	Minus
M	Missing data
M	Moderate echoes
M	Mountain Standard Time
MA	Map Analysis
MA	Massachusetts
MAN	Manitoba
MAR	At or over sea
MAX	Maximum

MD	Maryland
MDT	Moderate
ME	Maine
MEGG	Merging
METAR	Aviation routine weather report
MEX	Mexico
MHKVLY	Mohawk Valley
MI	Michigan
MI	Shallow
MID	Middle
MIDN	Midnight
MIFG	Patches of shallow fog
MISG	Missing
MLD	Mild
MLTLVL	Melting Level
MMO	Main Meteorological Office
MN	Minnesota
MNLD	Mainland
MNM	Minimum
MO	Missouri
MOD	Moderate

MOGR	Moderate Or Greater
MON	Above or over Mountains
MONTR	Monitor
MOV	Move
MOVG	Moving
MRGL	Marginal
MRNG	Morning
MRTM	Maritime
MS	Minus
MS	Mississippi
MSTLY	Mostly
MSTR	Moisture
MT	Maritime Tropic
MT	Montana
MTN	Mountain
MTNS	Mountains
MTW	Mountain Waves
MVFR	Marginal Visual Flight Rules
MXD	Mixed
N	No change
NB	New Brunswick

NC	No Change, Not Changing
NC	North Carolina
NCWX	No Change in Weather
ND	North Dakota
NE	Nebraska
NE	No Echoes
NELY	Northeasterly
NERN	Northeastern
NFLD	Newfoundland
NGT	Night
NH	New Hampshire
NJ	New Jersey
NL	No Layers
NLM	Nautical Mile
NM	New Mexico
NMBR	Number
NMRS	Numerous
NNERN	North-Northeastern
NNWRN	North-Northwestern
NNWWD	North-Northwestward
NO	Not available
NO	NOTAM

NORPI	No pilot balloon observation
NOSIG	Not Significant
NPRS	Nonpersistent
NR	No Report
NRN	Northern
NRW	Narrow
NS	Nimbostratus
NS	NOTAM Summary
NS	Nova Scotia
NSC	No Significant Cloud
NSCSWD	No Small Craft or Storm Warning Displayed
NSW	No Significant Weather
NV	Nevada
NVA	Negative Vorticity Advection
NWLY	Northwesterly
NWRN	Northwestern
NY	New York

OAOI	On And Off Instruments
OBS	Observe
OBSC	Obscure

OBSCD	Obscured
OCFNT	Occluded Front
OCL	Occlude
OCLD	Occlude
OCLN	Occlusion
OCN	Occasion
OCNL	Occasional
OCNLY	Occasionally
OCR	Occur
OCTY	Over City
OFP	Occluded Frontal Passage
OFSHR	Offshore
OH	Ohio
OI	On Instruments
OK	Oklahoma
OMTNS	Over Mountains
ONSHR	Onshore
ONT	Ontario
OR	Oregon
ORGPHC	Orographic
OSV	Ocean Station Vessel
OTAS	On Top And Smooth

OTLK	Outlook
OTP	On Top
OTR	Other
OVC	Overcast (8/8 cloud coverage)
OVHD	Overhead
OVR	Over
OVRNG	Overrunning
OWRSCTD	Lower Scattered
P	Pacific Standard Time
P	Polar
P	Probability
P-	Light Precipitation in unknown form
PA	Prefix for Alaska
PA	Pennsylvania
PAC	Pacific
PBL	Probable
Pc	Polar Continental
PCLL	Persistent Cell
PCPN	Precipitation
PDMT	Predominant

PMT	Prominent
PDW	Priority Delayed Weather
PE	Ice Pellets
PEN	Peninsula
PGTSND	Puget Sound
PH	Prefix for Hawaii
PIBAL	Pilot Balloon
PISE	No Pilot balloon report due to unfavorable Sea conditions
PISO	No Pilot balloon report due to Snow
PIWI	No Pilot balloon report due to high Winds, or gusty surface Winds
PK	Peak
PLATF	Plain Language Terminal Forecast
PLW	Plow(ed)
Pm	Polar Maritime
PNHDL	Panhandle
PO	Dust devils, sand whirls
PPINA	Radar weather report Not Available
PPINE	Radar weather report No Echoes

PPINO	Radar weather report Inoperative due to breakdown
PPIOK	Radar weather report equipment operation resumed
PPIOM	Radar weather report equipment Inoperative due to Maintenance
PR	Partial
PR	Pressure
PR	Puerto Rico
PRBLTY	Probability
PRD	Period
PRES	Pressure
PRESFR	Pressure Falling Rapidly
PRESRR	Pressure Rising Rapidly
PRIND	Present Indications
PRJMP	Pressure Jump
PROB	Probabilities
PROG	Prognosis, Prognostic, Progress
PRSNT	Present
PRST	Persist
PS	Plus
PSBL	Possible

PSG	Passage, Passing
PSN	Position
PTCHY	Patchy
PTLY	Partly
PTN	Portion
PVA	Positive Vorticity Advection
PVL	Prevail
PY	Spray
Q	International altimeter setting in millibars
Q	Squall
QSTNRY	Quasistationary
QUAD	Quadrant
QUADS	Quadrants
QUE	Quebec
R	Radar, Radiosonde (ceiling reading)
R	Rain
R	Right
R-	Light Rain

RA	Radiosonde
RA	Rain
RABA	No upper winds observation, no Balloons are Available
RABAL	Radiosonde Balloon wind data
RABAR	Radiosonde Balloon Release
RACO	No upper winds observation, Communications are Out
RADAT	Radiosonde observation Data (freezing level data)
RADNO	Report missing account radio failure
RAFI	Radiosonde observation not Filed
RAFRZ	Radiosonde observation Freezing levels
RAHE	No upper winds observation, no gas available
RAICG	Radiosonde observation, Icing at...
RAOB	Radiosonde Observation
RAREP	Radar weather Report
RAVU	Radiosonde Analysis and Verification Unit
RAWE	No upper winds observation, unfavorable Weather

RAWI	No upper winds observation, high and gusty Winds
RAWIN	Upper Winds observation
RB	Rain Began
RCD	Radar Cloud Detection report
RCDNA	Radar Cloud Detection report Not Available
RCDNE	Radar Cloud Detection report No Echoes observed
RCDNO	Radar Cloud Detector not operative due to breakdown until
RCDOM	Radar Cloud Detector inoperative due to Maintenance until
RCH	Reach
RCHG	Reaching
RCKY	Rocky
RDG	Ridge
RDGS	Ridges
RDWND	Radar Dome Wind
REDVLPG	Redeveloping
REFRMG	Reforming

RESTR	Restrict
RGD	Ragged
RGN	Region
RH	Relative Humidity
RHINO	Radar range Height Indicator Not Operating
RI	Rhode Island
RIOGD	Rio Grande
RMDR	Remainder
RMK	Remark(s)
RMN	Remain
RMRK	Remark
RNFL	Rainfall
RNG	Range
RNWY	Runway
ROBEPS	Radar Operating Below Prescribed Standard
RPD	Rapid
RPDLY	Rapidly
RPm	Returning Polar Maritime
RPT	Repeat
RR	Rising Rapidly

RS	Record Special
RSG	Rising
RTD	Routine Delayed weather
RTRD	Retard
RTRN	Return
RUF	Rough
RVR	River
RW	Rain showers
RWU	Rain showers Unknown

S	Snow
SA	Record observation
SA	Sand
SA	Sequenced Aviation weather report (message identifier)
SA	Surface Actuals
SASK	Saskatchewan
SB	Snow Began
SBSD	Subside
SC	South Carolina
SC	Stratocumulus

SCAN	Significant Changes and Notices to Airmen
SCOB	Scattered Clouds Or Better
SCSL	Standing Lenticular Stratocumulus
SCT	Scattered ($\frac{3}{8}$ to $\frac{4}{8}$ cloud coverage)
SCTD	Scattered
SCTR	Sector
SD	Radar weather report
SD	South Dakota
SDEP	Smoke layer estimated Deep
SEC	Section
SELS	Severe Local Storms
SELY	Southeasterly
SERN	Southeastern
SEV	Severe
SFERICS	Atmospherics
SG	Snow Grains
SGD	Solar-Geophysical Data
SH	Showers
SHFT	Shift
SHLW	Shallow
SHRTLY	Shortly

SHWR	Shower
SIERNEV	Sierra Nevada
SIGMET	Significant Meteorology
SIR	Snow and Ice on Runway
SKC	Sky Clear
SKED	Schedule
SLD	Solid
SLGT	Slight
SLGTLY	Slightly
SLP	Sea Level Pressure
SLP	Slope
SLR	Slush on Runway
SLT	Sleet
SLW	Slow
SMK	Smoke
SML	Small
SMTH	Smooth
SMWHT	Somewhat
SN	Snow
SNBNK	Snowbank
SNFLK	Snowflake
SNO	Snow

SNOINCR	Snow Depth Increase in past hour
SNRS	Sunrise
SNST	Sunset
SNW	Snow
SNWFL	Snowfall
SP	Snow Pellets
SP	Special
SP	Station Pressure
SPECI	Special
SPKL	Sprinkle
SPLNS	South Plains
SPRD	Spread
SPRDG	Spreading
SPRL	Spiral
SQ	Squall
SQAL	Squall
SQLN	Squall Line
SQLNS	Squall Lines
SS	Sandstorm
SSERN	South-Southeastern
SSEWD	South-Southeastward

SSWRN	South-Southwestern
SSWWD	South-Southwestward
ST	Stratus
STAGN	Stagnation
STBL	Stable
STDY	Steady
STFR	Stratus Fractus
STFRA	Stratus Fractus
STFRM	Stratiform
STG	Strong
STM	Storm
STN	Station
STNR	Stationary
STNRY	Stationary
SUNRS	Sunrise
SVR	Severe
SVRL	Several
SW	Snow showers
SWLG	Swelling
SWLY	Southwesterly
SX	Stability index
SXN	Section

SXNS	Sections
SYNOP	Synoptic
SYNS	Synopsis
SYS	System

T	Temperature
T	Thunderstorm
T	Trace
T	Tropical
T+	Severe Thunderstorm
T/D	Temperature/Dew point spread
TAF	Terminal Area Forecast
Tc	Tropical Continental
TCU	Towering Cumulus
TDA	Today
TDO	Tornado
TEFOR	Terminal aviation Forecast
TEMP	Temperature
TEMPO	Temporarily
THD	Thunderhead
THDR	Thunder

THK	Thick
THN	Thin
THRFTR	Thereafter
THRU	Through
THRUT	Throughout
THSD	Thousand
THTN	Threaten
TIL	Until
TL	Until
Tm	Tropical Maritime
TMP	Temperature
TMW	Tomorrow
TN	Tennessee
TNDCY	Tendency
TNGT	Tonight
TNO	Thunderstorm
TOVC	Top of Overcast
TPG	Topping
TRIB	Tributary
TRML	Terminal
TROF	Trough
TROP	Tropopause

TRPCD	Tropical Continental
TRPCL	Tropical
TRPLYR	Trapping Layer
TRRN	Terrain
TRS	Tropical cyclone
TS	Thunderstorm
TSHWR	Thundershower
TSQLS	Thundersqualls
TSTM	Thunderstorm
TT	Temperature
TURB	Turbulence
TURBC	Turbulence
TURBT	Turbulent
TWD	Toward
TWRG	Towering
TX	Texas

U	Unknown
U	Up
UA	Upper Air, pilot report (PIREP) (message identifier)
UAG	Upper Atmosphere Geophysics

UDDF	Updrafts and Downdrafts
UNK	Unknown
UNL	Unlimited
UNRSTD	Unrestricted
UNSBL	Unseasonable
UNSTBL	Unstable
UNSTDY	Unsteady
UNSTL	Unsettle
UP	Unknown Precipitation
UPDFTS	Updrafts
UPR	Upper
UPSLP	Upslope
UPSTRM	Upstream
UPWD	Upward
USP	Urgent Special Report
UT	Utah
UU	Highest relative humidity
UVV	Upward Vertical Velocity
UWNDS	Upper Winds
V	Variable
VA	Virginia

VA	Volcanic Ash
VAL	In Valleys
VC	Vicinity
VERVIS	Vertical Visibility
VI	Virgin Islands
VIS	Visibility
VLCTY	Velocity
VLNT	Violent
VLY	Valley
VR	Veer
VRB	Variable
VRBL	Variable
VRISL	Vancouver Island, B.C.
VRTMOTN	Vertical Motion
VSB	Visible
VSBY	Visibility
VSBYDR	Visibility Decreasing Rapidly
VSBYIR	Visibility Increasing Rapidly
VSP	Vertical Speed
VT	Vermont
VV	Vertical Visibility

VV	Visibility Value
VW	Very Weak echoes
VWS	Vertical Wind Shear
W	Indefinite ceiling
W	Showers
W	Warm
W	Weak echoes
W	Western
W+	Weak echoes increasing
W–	Weak echoes weakening
WA	Washington
WA	Weather Advisory, AIRMET (message identifier)
WDC-1	World Data Centers in Western Europe
WDC-2	World Data Centers throughout the rest of the world
WDLY	Widely
WDSPR	Widespread
WDSPRD	Widespread
WEA	Weather

WFP	Warm Front Passage
WH	Hurricane advisory
WI	Wisconsin
WINT	Winter
WK	Weak
WKN	Weaken
WL	Will
WLY	Westerly
WND	Wind
WNWRN	West-Northwestern
WNWWD	West-Northwestward
WPLTO	Western Plateau
WR	Wet Runway
WRM	Warm
WRMFNT	Warm Front
WRNG	Warning
WS	SIGMET (message identifier)
WS	Windshear
WSHFT	Wind Shift
WSOM	Weather Service Operations Manual
WSR	Wet Snow on Runway

WST	Convective SIGMET (message identifier)
WSWRN	West-Southwestern
WSWWD	West-Southwestward
WT	Weather Trends
WTR	Water on Runway
WTRS	Waters
WTSPT	Waterspout
WV	Wave
WV	West Virginia
WW	Severe Weather Watch bulletin
WW-A	Severe Weather Watch Amended
WWD	Westward
WX	Weather
WXCON	Weather reconnaissance flight pilot report
WY	Wyoming

X	Obscured ceiling, surface based obstruction
XCP	Except
XPC	Expect

XTND	Extend
XTRM	Extreme
Y	Yukon Standard Time
YKN	Yukon
YLSTN	Yellowstone
ZI	Zonal Index
ZI	Zone of Interior
ZL	Freezing drizzle
ZR	Freezing Rain
ZRNO	Freezing Rain information Not available

WORD ENDINGS FOR WEATHER WORDS

BL	-able
CLY	-cally
D	-ed, -ied
DR	-der
G	-ing
L	-al

LY	-ly, -ally
MT	-ment
NC	-ance, -ence
NG	-ening
NS	-ness
R	-er, -ier, -or
RN	-ern
RY	-ary, -ery, -ory
S	-es, -ies, -s
ST	-est
TY	-ity
US	-our
V	-ive
WD	-ation, -tion

RUNWAY VISUAL RANGE INITIALS

BC	Patches
BL	Blowing
BR	Mist
DR	Drifting
DS	Dust Storm
DU	Widespread Dust

DZ	Drizzle
FC	Funnel Cloud
FG	Fog
FU	Smoke
FZ	Super-cooled
GR	Hail
GS	Small hail
HZ	Haze
IC	Diamond dust
MI	Shallow
PE	Ice Pellets
PO	Well-developed dust/sand whirls
RA	Rain
SA	Sand
SG	Snow Grains
SH	Shower(s)
SN	Snow
SQ	Squalls
SS	Sandstorm
TS	Thunderstorm
VC	In the Vicinity

URGENT PILOT REPORT INITIALS

/IC	Icing
/RM	Remarks
/SK	Sky
TA	°C
/TB	Turbulence
/TM	Time Zulu
/TP	Type of aircraft
/WV	Wind
/WX	Weather

THREE

Codes, Symbols, Designations, and Initials

ENGLISH

'	Feet
"	Inches
+	Also, In addition to, Positive
−	Negative
1301	Halon fire bottle
3D	Three-Dimensional
5606	Hydraulic fluid of mineral composite
a	Acceleration
A	Amps
A	Area
A_j	Area of the Jet
C	Centrifugal force
°C	Celsius (formerly centigrade)
c	Sonic velocity
C_d	Coefficient of Drag
C_l	Coefficient of Lift

$C_{l\,(max)}$	Maximum Coefficient of Lift
C_m	Pitching Moment Coefficient
C_n	Yawing Moment Coefficient
C_p	Constant Pressure
C_s	Molecular Speed
C_u	Copper
C_v	Constant Volume
D	Diameter
D	Dip angle
d	Distance
D	Dynamic
D_h	Hydraulic Diameter
D_i	Induced Drag
D_p	Parasite Drag
D_t	Total Drag
E	Efficiency
F	Force
F	Thrust
°F	Fahrenheit
F_a	Molecular impact Force
FB	Braking Force
F_g	Total thrust

F_n	Net thrust
F_r	Ram drag
F_u	Net thrust effect
g	Acceleration of Gravity
H	Height
H	Horizontal component
H	Total pressure (psf)
H^f	Latent Heat of Fusion
Hg	Mercury
hPa	Hectopascal
I	Current
I	Inner
K	Constant
K	Number of cylinders
°K	Kelvin (273°+°C)
k	Thermal conductivity
KE	Kinetic Energy
L	Length
M	Mach
M	Mass
m	Molecular Mass
MHz	Megahertz

n	Exponent
N	Normal force
N	Number
nth	Thermal efficiency
o	Outer
p	Air density
P	Power
P	Pressure
P	Primary
P	Static Pressure
P_{am}	Absolute Barometric Pressure
PE	Potential Energy
P_j	Static pressure (absolute pressure)
P_o	Standard sea level static Pressure
Q	Airflow
q	Dynamic pressure
Q	Mass airflow
r	Radial distance
R	Resistance
R	Universal gas constant (absolute zero)
"R"	Gas constant

°R	Rankine temperature
Re	Reynolds number
S	Secondary
S	Static
S	Wing area
T	Temperature
T	Thrust
t	Time
T	Total force
T_o	Standard sea level temperature
T_t	Total Temperature
Tu	Turbulence intensity
V	Speed
V	Velocity Vector
V	Volts
V1	Initial Velocity
V2	Final Velocity
$V_{A/C}$	Vector of an Aircraft
V_h	Horizontal component Vector
V_K	Airspeed in Knots
V_R	Resultant Vector
V_V	Vertical component Vector

V_w Vector of Wind
W Gas Weight
W Watts
W Weight
W Work
W_a Airflow rate
W_a Weight of compressed Air
W_f Fuel Flow rate
W_f Weight of Fuel
Z Vertical component
Z_P Pressure altitude

GREEK

α (alpha) Angle of attack
Δ (delta) Change
δ (delta) Pressure Ratio
θ (theta) Bank angle, Temperature ratio
μ (mu) Coefficient of friction
ν (nu) Kinematic viscosity
Σ (sigma) Sum

SYMBOLS AND SHORTHAND

/	And
@	At
^	Celestial assumed position
≠	Cross
↑	Climb
↓	Descend
-	Direct
✓	From previous
⊗	Intersection
))	Log entry
#	Number, Pounds
=	Radial
×	Times

FOUR

Aviation Alphabet

A Alfa
B Bravo
C Charlie
D Delta
E Echo
F Foxtrot
G Golf
H Hotcl
I India
J Juliet
K Kilo
L Lima
M Mike
N November
O Oscar
P Papa
Q Quebec

R Romeo
S Sierra
T Tango
U Uniform
V Victor
W Whiskey
X X ray
Y Yankee
Z Zulu

FIVE

Slang and Memory Joggers

SLANG

Air Sense
Experienced pilot

Bought the Farm
Pilot crashed

Bug Smasher
Small training aircraft

Coffin Corner
Critical area of the aircraft's flight envelope

Dead Stick
Stopped engine

Hell Hole
Compartment of an aircraft where a major portion of the systems is located; this is not a favored work area of mechanics

MAYDAY
Distress call

Nonrev
Nonrevenue passenger—usually a free ticket passenger, such as an off-duty pilot or flight attendant

P-Factor
A physical pressure, located in the lower abdominal area, that should be relieved before a flight

PANPAN
Urgency call

Puddle Jumper
Small float plane

Red Eye Flight
Late night flight

SOBs
Souls on Board—the number of persons aboard the aircraft

St. Elmo's Fire
Static electrical display

MEMORY JOGGERS

3 Hs

Hot, **H**igh, and **H**umid
(hazardous take-off conditions)

5 Ts

Time, **T**urn, **T**wist, **T**hrottle, and **T**alk
(one of many versions of the Ts);
at any intersection, checkpoint, fix, or turn

Aviate, Navigate, then Communicate
Order of importance

East Is Least and West Is Best
Altitude and variation aid

PRELANDING CHECK

Gumps

G Gas
U Undercarriage (landing gear)
M Mixture
P Propeller
S Safety (belts...)

High to Low
Look out below

Low to High
Clear the sky

LOST COMMUNICATIONS PROCEDURES

A Assigned
V Vectored
E Expected
F Filed
M Minimum IFR altitude
E Expected
A Assigned

MULTIENGINE, SINGLE-ENGINE Vmc FACTORS

Little Charlie Won't Buy FAT COWS

L Landing gear up
C Cowl Flaps in take-off position
W Maximum gross Weight
B No *more* than 5° Bank into the good engine
F Flaps in take-off position
A At maximum power on good engine or METO
T Trim for take-off
C Center of gravity in most unfavorable position
O Out of ground effect
W Windmilling prop on inoperative engine
S Standard conditions at sea level

PERSONAL CHECKLIST

I'm Safe

I Illness
M Medical
S Stress
A Alcohol
F Fatigue
E Emotion

POSITION REPORT

PTA Ten

P Position
T Type (IFR)
A Altitude
T Time (current)
E Estimated time to next reporting point
N Next reporting point

WHEN NOT TO PERFORM A PROCEDURE TURN

Sharp-T

S Straight-in
H Holding pattern
A Arc (DME)
R Radar vectoring
P Procedure turn (no PT)
T Timed approach

VOR CHECK REQUIREMENTS TO RECORD

Deps

D Date
E Error
P Place
S Signature

STEPS TO PERFORM ON A MISSED APPROACH

Pitch Up
Power Up
Clean Up
Talk Up

WEATHER FRONTS IN ORDER OF SEVERITY

Cows

C Cold
O Occluded
W Warm
S Stationary

COMPASS ERRORS

Ands

Accelerate **N**orth
Decelerate **S**outh

WEIGHT AND BALANCE AID

Wam

Weight × **A**rm = **M**oment

APPROACH BRIEFING

Ice ATM

I Identify the station
C Course to be flown
E Entry
A Altitude
T Time
M Missed approach procedures

AIRSPEED ORDER

Ice-T

I Indicated
C Calibrated
E Equivalent
T True

123 RULE FOR FILING AN ALTERNATE

1 1 hour before or after
2 2000-ft ceiling
3 3 mi visibility

IFR CLEARANCE LIST

Craft

C Clearance Limit
R Route
A Altitude
F Frequency
T Transponder code

HEADING ORDER

Can Dead Men Vote Twice?

C Compass
D Deviation
M Magnetic
V Variation
T True

or

True Virginians Make Delightful Company

T True
V Variation
M Magnetic
D Deviation
C Compass

PREMANEUVER CHECKLIST

Stop

S Safe
(airplane, area to land, area-clearing turns)

T Tolerances
(heading, altitude, airspeed, bank angle)

O Objective
(smoothness, accuracy, division of attention)

P Procedure
(coordination, orientation, planning)

OBJECTIVE AND PROCEDURE

Sad Cop

S Smoothness
A Accuracy
D Division of attention
C Coordination
O Orientation
P Planning

VFR DAY REQUIRED EQUIPMENT

Tomato Flames

T Tachometer
O Oil pressure gauge
M Manifold absolute pressure
A Altimeter
T Temperature gauge—liquid cooled
O Oil temperature—air cooled

F Fuel gauge
L Landing gear position indicator
A Airspeed indicator
M Magnetic direction indicator
E Emergency locator transmitter
S Seat belts

VFR NIGHT REQUIRED EQUIPMENT

Flaps

F Spare Fuses
L Landing light
A Anticollision
P Position lights
S Source of electric energy

WEATHER AID

6, 5, 4, 3, 2, 5

WAs are good for **6 hours** and contain **5 items.**
WSs are good for **4 hours** and contain **3 items.**
WSTs are good for **2 hours** and contain **5 items.**

REQUIRED INSPECTIONS

Haste

H Hundred hour (commercial)
A Annual
S Static/pitot
T Transponder
E Emergency locator transmitter

REPORT TO ATC

I Paids

I Identification
P Position
A Altitude
I Intentions or
D Destination
S Squawk

SPIN RECOVERY TECHNIQUE

Parent

P Power
A Ailerons
R Rudder
E Elevator
N Neutralize
T Trim

USELESS TO PILOTS

Sky Above
Runway behind
Fuel burned (or left in the fuel pump)

AIRCRAFT DOCUMENTS

Arrow

A Airworthiness certificate
R Registration certificate
R Radio station license
O Operating limitations
W Weight and balance data

CFI HELPERS

LAWS OF LEARNING

Reepir

R Recency
E Exercise
E Effect
P Primacy
I Intensity
R Readiness

PRIVATE PILOT CHECKLIST OF SIGN-OFFS

5, 4, 3, 2, 1

5 Solo endorsements
4 Cross-country endorsements
3 Check ride recommendations
2 Class "B" endorsements
1 Written

SIX

V-Speeds

V1	Take-off decision speed
V2	Take-off safety speed (minimum)
Va	Design maneuvering speed
Vb	Design speed for maximum gust
Vc	Design Cruising speed
Vd	Design Diving speed
Vdf/Mdf	Demonstrated Flight diving speed
Vf	Design Flap speed
Vfc/Mfc	Maximum speed for stability Characteristics
Vfe	Maximum Flap Extended speed
Vfo	Maximum Flap Operating speed
Vh	Maximum speed in level flight with maximum continuous power
Vle	Maximum Landing gear Extended speed
Vlo	Maximum Landing gear Operating speed

Vlof	Liftoff speed
Vmc	Minimum Control speed
Vmca	Air Minimum Control speed
Vmcg	Ground Minimum Control speed
Vmo/Mmo	Mach, Maximum Operating limit speed
Vmu	Minimum Unstick speed (liftoff)
Vne	Never Exceed speed
Vno	Maximum structural crusing speed
Vpw	Pilot Window open speed
Vr	Rotation speed
Vref	Reference speed
Vs	Stalling speed
Vs1	Stalling speed in a specified configuration
Vso	Stalling speed in the landing configuration
Vsse	Safe Single-Engine speed (render an engine inop)
Vtos	Take-Off Safety speed
Vtoss	Take-Off Safety speed category "A" rotorcraft

V-Speeds

Vww Windshield Wiper operating speed

Vx Best angle of climb speed

Vxse Best angle of climb speed, Single Engine

Vy Best rate of climb speed

Vyse Best rate of climb speed, Single Engine

SEVEN

The Military and Manufacturers

MILITARY DESIGNATIONS

#(4)	Fourth aircraft (# from the manufacturer)
A	Attack
B	Bomber
C	Cargo
D	Director
E	Electronic
F	Fighter
G	Glider
H	Helicopter
K	Refueler
L	Liaison
M	Multimission
O	Observation
P	Pursuit
Q	Drone

R Reconnaissance

S Submarine hunter

T Trainer

U Utility

V VIP transport/Vertical or STOL

X Experimental

Y Prototype

Z Lighter than air

MANUFACTURER DESIGNATIONS

A Brewster

B Boeing

C Curtiss

D Douglas

E Piper

F Grumman

J North America

K Fairchild

L Lockheed

M Martin

P Northrop

S Stearman

U Chance Vought

V Vultee

W Waco

Y Consolidated

MANUFACTURER ABBREVIATIONS

A	Airbus
An	Antonov
ATR	Aerospatiale Transport Regional
B	Boeing
BAe	British Aerospace
BN	Pilatus Britten Norman
C	Cessna
Concorde	Aerospatiale British Aerospace
Dauphin	Eurocopter
DC	McDonnell Douglas
DHC	de Havilland Canada
Do	Dornier
EMB	Embraer
F	Fokker
IL	Ilyushin
Ka	Kamov

L	Lockheed
MD	McDonnell Douglas
Mercure	Dassault
Mi	Mil
PA	Piper
SD	Short Brothers
Tu	Tupolev
Yak	Yakovlev

EIGHT

Transponder Codes and Squawk Codes

TRANSPONDER CODES

/A DME, transponder mode C

/B DME, transponder mode A

/C RNAV, transponder mode A

/D DME, no transponder

/E FMS and VNAV with oceanic, en route, and terminal navigation and approach capabilities

/F FMS and VNAV with en route and terminal navigation and approach capabilities

/G Flight management system and electronic flight management system with **/R** capability (GPS)

/M TACAN, no transponder

/N TACAN, transponder mode A

/P TACAN, transponder mode C

/R RNAV, transponder mode C

/T transponder mode A

/U transponder mode C

/W RNAV, no transponder

/X No transponder

SQUAWK CODES

7777 Military, Interceptor

7700 Emergency

7600 Lost communications

7500 Hijack

1200 Visual flight rules

About the Author

Christopher J. Abbe (Saginaw, MI) is an airline pilot with a B.S. in aviation technology and operations. Abbe has been flying since he soloed on his sixteenth birthday. He is a Gold Seal (CFI, CFII, MEI, AGI) flight instructor.